4th International Conference on Manufacturing and Industrial Technologies (ICMIT 2017)

IOP Conference Series: Materials Science and Engineering
Volume 212

Lisbon, Portugal
27 – 29 May 2017

ISBN: 978-1-5108-4414-8
ISSN: 1757-8981

Printed from e-media with permission by:

Curran Associates, Inc.
57 Morehouse Lane
Red Hook, NY 12571

Some format issues inherent in the e-media version may also appear in this print version.

Printed by Curran Associates, Inc. (2017)

For permission requests, please contact the Institute of Physics
at the address below.

Institute of Physics
Dirac House, Temple Back
Bristol BS1 6BE UK

Phone: 44 1 17 929 7481
Fax: 44 1 17 920 0979

techtracking@iop.org

Additional copies of this publication are available from:

Curran Associates, Inc.
57 Morehouse Lane
Red Hook, NY 12571 USA
Phone: 845-758-0400
Fax: 845-758-2633
Email: curran@proceedings.com
Web: www.proceedings.com

Table of contents

Volume 212

The 4th International Conference on Manufacturing and Industrial Technologies
27–29 May 2017, Lisbon, Portugal

Accepted papers received: 14 June 2017
Published online: 29 June 2017

Preface

Automobile design and manufacture

Mechanical design-manufacture and automation

Industrial production and safety management

Power and energy systems

IOP Conf. Series: Materials Science and Engineering **212** (2017) 011001 doi:10.1088/1757-899X/212/1/011001

PREFACE

2017 4th International Conference on Manufacturing and Industrial Technologies (ICMIT 2017) was held in Lisbon, Portugal during May 27-29, 2017. ICMIT 2017 is dedicated to issues related to Manufacturing and Industrial Technologies.

The major goal and feature of the conference is to bring academic scientists, engineers, industry researchers together to exchange and share their experiences and research results, and discuss the practical challenges encountered and the solutions adopted. Professors from Portugal, Turkey and Romania were invited to deliver keynote speeches and plenary speech regarding latest information in their respective expertise areas.

This proceedings present a selection from papers submitted to the conference from universities, research institutes and industries. All of the papers were subjected to peer-review by conference committee members and international reviewers. The papers selected depended on their quality and their relevancy to the conference.

We are grateful for the submissions of the authors. We also would like to express our sincere gratitude to our colleagues in the disciplines who have kindly volunteered in the review process. And special thanks go to the organizers and chairs of the conference and the sessions.

The various panels that have been organized during ICMIT 2017 promised exciting revelations for all who attended them. ICMIT 2017 is the background on which every participant might learn more about a topic with which one is already familiar; or might also find oneself discovering a whole new world of ideas and information one didn't know existed. It is our hope that the proceedings of ICMIT 2017 will help to find new ideas in different areas.

Prof.Mário S. Ming Kong
University Lisbon, Portugal
June 3th, 2017

IOP Conf. Series: Materials Science and Engineering **212** (2017) 011002 doi:10.1088/1757-899X/212/1/011002

Conference Committee

Conference Chair

Prof. Mário S. Ming Kong, University Lisbon, Portugal

Program Chair

Dr.Andrea Deaconescu, Transilvania University of Brasov, Romania

Dr.İlhan Asiltürk, Selcuk University, Turkey

Dr.Serhat İkizoğlu, Istanbul Technical University, Turkey

Dr.Osman Adiguzel, Firat University, Department of Physics, Turkey

Dr. Wen.Z. Cui, Chongqing University, China

Technical Committee

Prof. João C. O. Matias, University of Aveiro, Portugal

Dr. Emin Taner ELMAS,Forent Energy Co., İzmir / TURKEY

Dr. Mehdi Jafari Vardanjani, University of Tehran, Iran

Dr. Ahmed Maher Mohammed, University of Portsmouth, UK

Dr. Wan Ahmad Yusmawiza Wan Yusoff, Hail University, Kingdom of Saudi Arabia

Dr. Ergin Kilic, Suleyman Demirel University, Turkey

Dr. Alireza Araee, University of Tehran, Iran

Dr. Erik Prada, Slovak joint stock research company,Slovak

Dr. Eng. Ivan Virgala,Technical University of Košice,Slovak

Dr. Chien-Chih Wang,Ming Chi University of Technology, Taiwan

Dr. Kuentai Chen, Ming Chi University of Technology, Taiwan

Dr. K. K. Soundrapandian, Pdpm-Indian Institute Of Information Technology Design And Manufacturing, Jabalpur, India

Dr. Tzu-An Chiang, National Taipei University of Business, Taiwan

Dr. Mohd Omar bin Ab Kadir, Universiti Sains Malaysia (Usm), Malaysia

Dr. Noor Aziah Serri, Universiti Sains Malaysia (Usm), Malaysia

IOP Conf. Series: Materials Science and Engineering **212** (2017) 011003 doi:10.1088/1757-899X/212/1/011003

Peer review statement

All papers published in this volume of *IOP Conference Series: Materials Science and Engineering* have been peer reviewed through processes administered by the proceedings Editors. Reviews were conducted by expert referees to the professional and scientific standards expected of a proceedings journal published by IOP Publishing.

IOP Conf. Series: Materials Science and Engineering **212** (2017) 012001 doi:10.1088/1757-899X/212/1/012001

Identifying Technical Condition of Vehicle Gearbox Using Acoustic Emission

Jan Furch, Josef Glos

Kounicova 65, 662 10 Brno, Czech Republic

jan.furch@unob.cz

Abstract. The article examines the technical condition of rotating parts using acoustic diagnostics. The measured object was a mechanical transmission of a field vehicle. Recently this method has been developing very quickly and is expected to be used not only for the signal analysis itself, but also for the failure occurrence prediction which is our aim in the future. In our article we observe the technical condition of a four-speed transmission and analyse the acoustic signal expressed by the root mean square of a noise level in decibels.

1. Introduction

Acoustic emission (AE) has proven to be an effective non-destructive technique to investigate the behaviour of material under mechanical stress. Compared with vibration techniques, AE offers several advantages. For example, AE techniques are capable of incipient fault detection. Additionally, it is sensitive to fault location, allowing its use for fault location detection. When applied to rolling element bearings, it has been shown that AE techniques can detect faults earlier than other technologies [1].

In the paper we present advanced computational models suitable for the development of a modern powertrain in the field of noise and vibration. The aim is to decide how detailed the model should be to correctly describe the vibrational and acoustic performance of the powertrain [2].

A lot of companies spend big amounts of money on the preventive maintenance of machines and technical equipment despite the fact that at that particular moment there is no need to take a preventive action. This leads to big financial losses, mainly during the time the machine is unavailable. Therefore, it is desirable to perform preventive maintenance at the time when it is really necessary. Determining a preventive action or replacing a part based on the real technical condition with the use of a mathematical model describing the machine technical condition during its life leads to a proactive maintenance. The aim of our paper is to describe the technical condition of a vehicle gearbox using acoustic emission.

2. Possibilities of detecting the technical condition of vehicle gearbox

Noise as a contactless measurable diagnostic value might be used as a source of information about a device technical condition. The noise is made by frictional forces and microscopic power impulses generated by mutual movement of rough or inaccurately smoothed surfaces at the point of a pair of parts contact. The disadvantage of the noise diagnostics is the existence of spurious reflections and interference in enclosed space.

Main characteristics of vibrations in acoustic emission

Acoustic amplitude y of a harmonic wave for spreading in the x direction is expressed by the following equation [3]

$$y = A\sin\omega\left(t - \frac{x}{c}\right) \quad (1)$$

for acoustic velocity v of a harmonic wave for spreading in the x direction it is expressed by [3]

$$v = A\omega\cos\omega\left(t - \frac{x}{c}\right), \quad (2)$$

where A – amplitude, c – wave spreading velocity, ω – angular velocity.

Acoustic pressure p during the wave passage is expressed by the equation [3]

$$p = \rho c v = p_A \cos\omega\left(t - \frac{x}{c}\right) = A\omega\rho c\cos\omega\left(t - \frac{x}{c}\right), \quad (3)$$

where p_A – acoustic pressure amplitude, ρ – environment density.

Sound rate I is a vector quantity usually defined as the mean value of specific sound power [3]

$$I = \frac{\mathrm{d}P}{\mathrm{d}S}, \quad (4)$$

where sound power P is determined by [3]

$$P = pSv, \quad (5)$$

where S is the plane perpendicular to the direction of waves spread.

The acoustic values are usually expressed in the levels related to referential values expressed in decibels. Therefore, for the introduced values the following equations apply [3]:

- sound power level L_W, where $P_0 = 10^{-12}$ W,

$$L_W = 10\log\frac{P}{P_0}; \quad (6)$$

- sound pressure level L_p, where $p_0 = 2\cdot10^{-5}$ Pa,

$$L_p = 20\log\frac{p_{ef}}{p_0} = 10\log\frac{p_{ef}^2}{p_0^2}; \quad (7)$$

- sound rate level L_I, where $I_0 = 10^{-12}$ Wm^{-2},

$$L_I = 10\log\frac{I}{I_0}. \quad (8)$$

The root of mean square (RMS) for a discrete sampled signal is defined as [4]:

$$RMS = \sqrt{\frac{1}{N}\sum_{i=1}^{N}(x_i^{\,2})} \quad (9)$$

where X_{RMS} is the root mean square value of data set x, x_i is the i-th element of x, N is the length of data set x.

The course of the signal usually depends on a time domain. The assessment of the signal time course is well completed by its properties represented by decomposition into the set of basic functions. The most natural approach for technical applications not only in the area of mechanical systems oscillation is the decomposition into the set of harmonic functions which differ in amplitude, angular frequency, and their initial phase. If the dependence of the amplitude and the initial phase on the frequency is illustrated for the set of harmonic signals, the signal is then illustrated in the frequency area – domain.

3. Description of approach to detecting a technical condition of vehicle gearbox

The noise of the gearbox is made by the interaction of its parts. For numerous mechanisms these interactions are random, disarranged impulses of the same size which cause in some parts of a machine being in a failure-free condition the oscillation of a broad and relatively steady spectrum.

3.1. Description of approach to detecting a technical condition of a measured bearing

Most of the damages to roller bearings occur because of wear. As a result, a fatigue of single components occur which leads to chipping out surface layer material (pitting), increasing wear and clearance, corroding process, grooving, deepening, and snap ring failures. Every time one of the rolling elements is damaged in any way, it starts to pulse, thereby stimulating natural frequencies in bearing rings. The speed of repetitive pulses causes a failure frequency which enables us to locate the damage to the roller bearing. The calculation of the failure frequencies are in Eq. (10) – (13). For this it is necessary to know the type of a bearing and its production code. The presented technique is validated using the acoustic emission signals of seeded fault steel bearings on a bearing test rig. The result is an effective acoustic emission–based approach validated to diagnose all four fault types: inner race, outer race, ball, and cage [5].

For the kinematic frequencies of impulses, provided the motion is all rolling, the formulae [6] apply:

BPFO - Ball Pass Frequency – Outer Race (outer ring defect)

$$f = \frac{n}{2} f_r \left(1 - \frac{BD}{PD}\cos\beta\right) \tag{10}$$

BPFI - Ball Passing Frequency Inner Race (inner ring defect)

$$f = \frac{n}{2} f_r \left(1 + \frac{BD}{PD}\cos\beta\right) \tag{11}$$

BSF - Ball Spin Frequency (bearing defect – of a bearing ball or a bearing roller)

$$f = \frac{PD}{2BD} f_r \left(1 - \left(\frac{BD}{PD}\cos\beta\right)^2\right) \tag{12}$$

FTF - Fundamental Train Frequency (snap ring defect)

$$f = \frac{1}{2} f_r \left(1 - \frac{BD}{PD}\cos\beta\right) \tag{13}$$

where n - the number of bearing balls or rollers, f_r - frequency given by the relative revolution (speed) of the inner and the outer ring, BP - the diameter of a bearing ball or roller, PD - pitch diameter, β-contact angle.

3.2. Description of approach to detecting a technical condition of measured gear

Vibrations caused by gear wheel engagement are transferred to outer surface of the gearbox and are changed into radiated noise.

The first source of vibrations is related to the principle of a gear pair. Changeable stiffness of mating involute gear wheels is the subject of numerous technical treaties. Many analytical studies are based on a single-mass model presenting flexible coupling of two mass gear wheels. After we had adapted two former dynamic model equations, we obtained only one dynamic equation of the second kind for the deviation *p(t)* which represents the above mentioned mutual movement towards the tangent of contact circles of both reduced mass gear wheels m_R, see [7]

$$m_R p(t) + B_R p(t) + K(t)(p(t) + \varepsilon(t)) = F(t). \quad (14)$$

The structure of the equation (14) conforms to the dynamic equation in which one parameter, stiffness, *K(t)* is the function of time. The outer excitation caused by vibration is represented by a dynamic force *F(t)* and a kinematic element $\varepsilon(t)$ given by deviations from the evolvent.

The second source of vibration excitation are the shocks when the teeth enter into engagement. These shocks might be also modelled by the step change in the teeth contact stiffness. The shocks are caused by different irregularities of teeth shape due to manufacturing deviations or deformation caused by teeth load. The important information for the diagnostics of teeth engagement is that the basic frequency of the stiffness or shocks change when the teeth enter into engagement. The frequency is given by the product of the frequency of toothed gear revolution and the number of its teeth. This frequency is called an engagement or tooth frequency. Since the stiffness changes according to the sinusoid, also the components with basic frequency multiples, i.e. its harmonic components, appear in the spectrum. Basic engagement frequency might be expressed as follows [7]

$$f_1 = z|f_0| \quad (15)$$

where: z – the number of teeth, f_0 – the frequency of toothed gear revolutions.
The harmonic components have a frequency f_k for which the following equation applies [7]:

$$f_k = kf_1, \text{ where } k > 1 \text{ is a natural number.} \quad (16)$$

4. Experiment description

The measurement was carried out on the laboratory stand which is shown in Fig. 1.

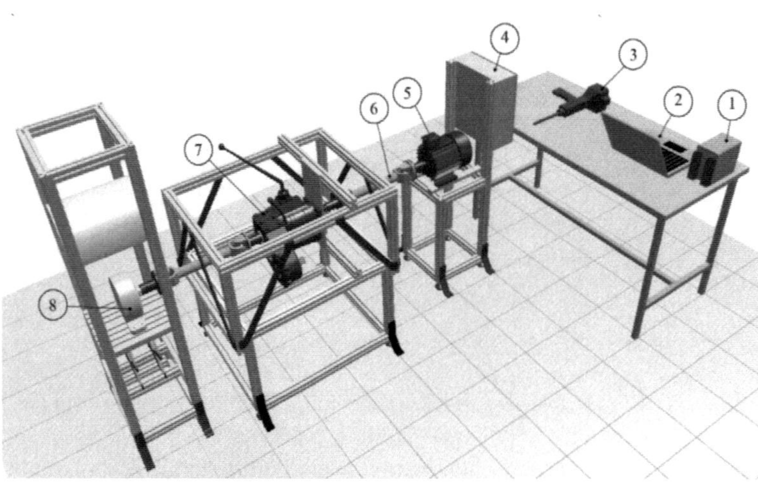

1 - Control unit for break AHB12;
2 - Laptop with DEWE software;
3 – The Ultraprobe 15000 kit;
4 - Speed control panel;
5 - Electric motor;
6 - Cardan shaft;
7 - The gearbox with transfer box;
8 - Air-cooled brake AHB-12.

Figure 1. Testing stand for acoustic measurement – 3D model.

In order to observe the changing gear technical condition, we have selected engine input speed of 2000 rpm. These input rpm cause a basic harmonic frequency of 33.3 Hz, or their multiples. Another selected parameter is shifting into third gear in the main gearbox, and "road" mode in the auxiliary gearbox. The mechanical changing gear was regularly observed up to the moment when the toothed gear of the third gear placed on the line shaft of the main gearbox got damaged, see Fig. 2.

Calculation of gearbox roller bearing faults

In order to find out the technical condition of the main gearbox we first calculated the frequencies of single roller bearings faults using the equations (10) – (13). In the gearbox there are the following roller bearings SKF6305, SKF 6208, SKF 6203, SKF3207BN, SKF6307, and SKF 6306N.

According to the location of roller bearings in the main gearbox, we calculated the frequencies of roller bearings faults based on single shafts revolutions, Table 1.

Table 1. Calculated values of the frequencies of roller bearings faults at input revolutions of 2000 rpm.

Location and specification of a roller bearing	Roller bearing revolutions [RPM]	BPFO [Hz]	BPFI [Hz]	BSF [Hz]	FTF [Hz]
SKF6208	2000	119.243	180.757	77.863	13.249
SKF3207BN	1266	77.871	112.029	49.151	8.652
SKF6307	1266	64.593	104.207	42.479	8.074
SKF6306N	938	47.736	77.331	31.182	5.967
SKF6305	938	40.253	69.180	27.505	5.750

After, we calculated the values of tooth frequencies in the main gearbox based on the input values mentioned earlier. When analysing the measuring results we focus on tooth revolutions occurring during the third gear shift which is in our case 500 Hz from the toothed gear of constant mesh transmission on a drive shaft and 421.88 Hz from the back toothed gear of the third gear shift, see Table 2. Other tooth frequencies are supressed since the single toothed gears, even if they are in mesh, are not loaded. During the observation of the gearbox technical condition we did not find any undesirable acoustic vibrations which would identify failure occurrence or a bearing apart from a gear toothing failure in the third gear shift.

Table 2.Calculated values of frequency of gearbox toothed gears depending on input revolutions

Pos.	Name of toothed gears	The number of teeth	Basic combustion engine harmonic frequency at 2000 rpm [Hz]	Tooth frequency at 2000 rpm [Hz]
1	Drive shaft and constant mesh gear	15	33.33	500
20	Constant mesh gear on line shaft	32	15.63	
17	Back gear 1st speed gear	15	15.63	234.38
7	Driven gear 1st speed gear	29	8.08	
18	Back gear 2nd speed gear	21	15.63	328.13
5	Driven gear 2nd speed gear	26	12.62	
19	**Back gear 3rd speed gear**	**27**	**15.63**	**421.88**
4	**Driven back gear 3rd speed gear**	**20**	**21.09**	
23 a	Idler reverse gear	19	12.34	234.38
17	Back gear 1st speed gear (for back action)	15	15.63	
23 b	Idler reverse gear	15	12.34	185.1
7	Driven gear 1st speed gear (for back action)	29	6.38	
10	Toothed gears for front axle drive– road drive	34	21.09	421.88
13		27	26.55	
14		34	21.09	

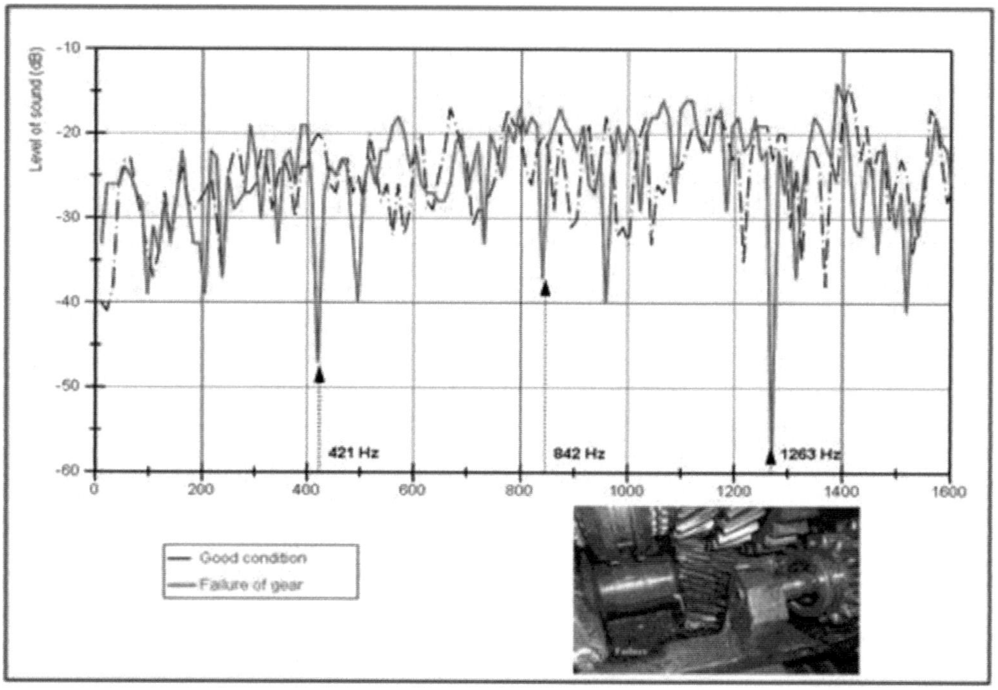

Figure 2. Expression of the dependence of gearbox technical condition on the noise level and the frequencies calculated from FFT.

5. Conclusion

The presented article is expected to inform the reader about the issue of acoustic emission and the diagnostics of engineering systems. We dealt with a mechanical four-speed gearbox with a two-speed auxiliary gearbox. As for the mechanical gearbox, we focused on the analysis of an acoustic signal created by roller bearings and single toothed gears. In the experiment we performed the measurement while shifting into third gear in the main gearbox and into "road" mode in the auxiliary gearbox. During the experiment a clearly identified failure occurred on the too thing of the third gear shift, see Fig. 2. Other parts of the gearbox did not show any signs of abnormal wear. The applied diagnostic method is a convenient tool for better identification of a mechanical failure in a transmission system.

References

[1] Mba, D. and Rao, R. Development of acoustic emission technology for condition monitoring and diagnosis of rotating machines; bearings, pumps, gearboxes, engines and rotating structures. The Shock and Vibration Digest, 2006. vol. 38, no. 1, pp. 3 – 16.

[2] NOVOTNY, Pavel, et al. Investigating the influence of computational model complexity on noise and vibration modeling of powertrain. *Journal of Vibroengineering*, 2016, 18.1.

[3] FAHY, F.J. Sound intensity. 2nd ed. London: E & FN Spon, 1995. ISBN 0419198105.

[4] QU, Yongzhi, et al. A new acoustic emission sensor based gear fault detection approach. IJPHM Special Issue on Wind Turbine PHM (Color), 2013, 32.

[5] VAN HECKE, Brandon; QU, Yongzhi; HE, David. Bearing fault diagnosis based on a new acoustic emission sensor technique. Proceedings of the Institution of Mechanical Engineers, Part O: Journal of Risk and Reliability, 2015, 229.2: 105-118.

[6] RENWICK, John T.; BABSON, Paul E. Vibration Analysis---A Proven Technique as a Predictive Maintenance Tool. IEEE Transactions on Industry Applications, 1985, 2: 324-332.

[7] KAHRAMAN, A.; SINGH, R. Interactions between time-varying mesh stiffness and clearance non-linearities in a geared system. Journal of Sound and Vibration, 1991, 146.1: 135-156.

IOP Conf. Series: Materials Science and Engineering **212** (2017) 012002 doi:10.1088/1757-899X/212/1/012002

Virtual Engine a Tool for Truck Reliability Increase

Jiri Stodola, Pavel Novotny

University of Defence Brno, Kounicova Str. 65, 662 10 Brno, Czech Republic
Brno University of Technology, Technicka Str. 2896/2, 616 69 Brno, Czech Republic

jiri.stodola@unob.cz

Abstract. The internal combustion engine development process requires CAD models which deliver results for the concept phase at a very early stage and which can be further detailed on the same program platform as the development process progresses. The vibratory and acoustic behaviour of the powertrain is highly complex, consisting of many components that are subject to loads that vary greatly in magnitude and which operate at a wide range of speeds. The interaction of the crank and crankcase is a major problem for powertrain designers when optimising the vibration and noise characteristics of the powertrain. The Finite Element Method (FEM) and Multi-Body Systems (MBS) are suitable for the creation of 3-D calculation models. Non-contact measurements make it possible to verify complex calculation models. All numerical simulations and measurements are performed on a Diesel six-cylinder in-line engine.

1. Introduction

In the powertrain development process several numeric simulation techniques have steadily gained in importance. The main target is design support; that is help for design decisions. With the increasing power of computers, accurate pre-calculations have become possible taking into account large number of effects and backward influences. First of all CAD models of Diesel in-line 6-cylinder engine are created. CAD models give basic geometric information about engine parts and can be used for a creation of FE models (e.g. crankshafts or engine blocks) or MBS rigid parts (e.g. pistons or piston pins). In the case of dynamic structural calculations the Finite Element Method (FEM) is 'state of the art'. The calculation of the structural transfer behaviour of single components is efficiently possible in the frequency domain, giving consideration to a large number of degrees of freedom. Figure 1shows FE models of main parts created on the basis of CAD models. The FE models of crankshafts or engine blocks are created in ANSYS. All FE models have uniform hexa element mesh and with exception in-line 6-cylinder crankshaft do not include small geometric details, for example, a radius on main bearing pin of a crankshaft. The uniform element size is suitable for dynamic solutions in time domain. The in-line 6-cylinder crankshaft includes small geometric details and is suitable for a stress-strain analysis.

2. Reduction of the models

Modal analyses can be performed in ANSYS and are not time-consuming, but for solution in time domain these models are very large and require reduction. The discretization of a flexible component into a finite element model represents the infinite number of DOF with a finite, but very large number of finite DOF [1]. The linear deformations of the nodes of this finite element model can be approximated as a linear combination of a smaller number of shape vectors, see equation (1).

IOP Conf. Series: Materials Science and Engineering **212** (2017) 012002 doi:10.1088/1757-899X/212/1/012002

$$u = \sum_{i=1}^{M} \Phi_i q_i \tag{1}$$

where
M ... number of mode shapes, Φ_i ... mode shape, q_i ... modal co-ordinates.

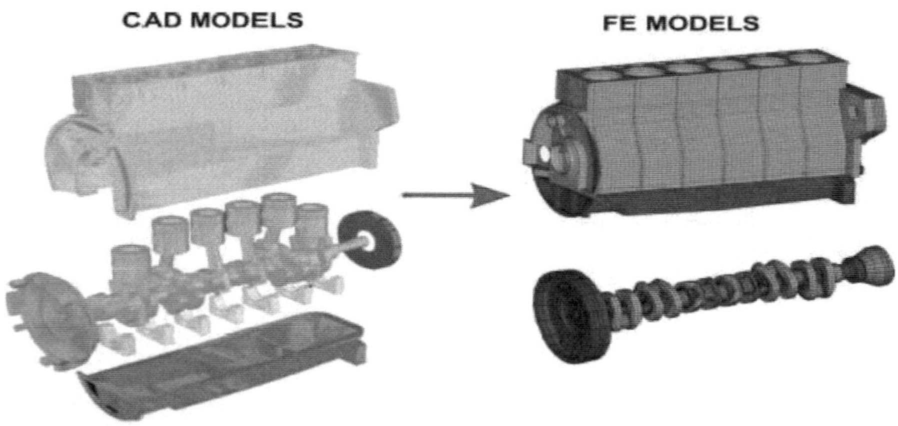

Figure 1. Diesel in-line 6-cylinder engine CAD and FE models

For reduction of FE models the Craig-Bampton method is used. The Craig-Bampton method separates the system DOF into boundary DOF, u_B and interior DOF, u_I . Two sets of mode shapes are defined as follows [2]:Constrain modes: these modes are static shapes obtained by giving each boundary DOF a unit displacement while holding all other boundary DOF fixed or Fixed-boundary normal modes: these modes are obtained by fixing the boundary DOF and computing an eigenvalue solution. The relationship between the physical DOF and Craig-Bampton modes and their modal co-ordinates is illustrated by the following equation:

$$u = \begin{matrix} u_B \\ u_I \end{matrix} = \begin{vmatrix} I & 0 \\ \varphi_{IC} & \varphi_{IN} \end{vmatrix} = \begin{matrix} q_C \\ q_N \end{matrix} \tag{2}$$

where
u_B.... column vector of boundary DOF, u_I .... column vector of interior DOF, I , 0 .... are identity and zero matrices, ϕ_{IC} .... physical displacements of interior DOF in the constrain modes, ϕ_{IN} .... physical displacements of interior DOF in normal modes, q_C .... column vector of the modal co-ordinates of the constrain modes, q_N .... column vector of the modal co-ordinates of the fixed-boundary normal modes.
The generalized stiffness and mass matrices corresponding to the Craig-Bampton modal basis are obtained via a modal transformation [3]. The orthogonalized Craig-Bampton modes are not eigenvectors of the original system. They are eigenvectors of the Craig-Bampton representation of the system and as such they have natural frequencies associated with them. The FE models of crankshafts and engine blocks are reduced to the number of boundary nodes selected by the user [3]. Their number is selected according to the application of constrains of the model. Loads are applied on these nodes only. The reduction of FE models from ANSYS into ADAMS is executed from ANSYS [1]. The parameters which influence the accuracy of the reduction are the number and location of the boundary nodes and the number of first modes considered during the reduction. These parameters can be

validated, for example, by computing FE model with constrains in ANSYS and separately in ADAMS and by subsequent comparison of the results.

3. Interaction between reduced Models

Slide bearings are used in the case of 6-cylinder in-line Diesel engine. The numerical solution of the equation of the motion in time domain with integration time step corresponding crank angular displacement 0.01° requires effective calculation of the hydrodynamic reaction of a radial slide bearing. The procedure derives from the Reynolds hydrodynamic equation [2]. The solution of hydrodynamic radial and axial slide bearings is implemented in ADAMS, where loads are obtained from a precalculated oil film reaction force database [3]. The Reynolds differential equation cannot be solved analytically and some numerical methods must be used. When hydrodynamic bearings are modelled, it can choose to include or exclude misalignment effects by using the following methods:

- For the two-dimensional method neglecting misalignment - an empirical analytical equation is used. This approach, which is similar to the impedance method, is the most efficient way to model hydrodynamic bearings.
- For the three-dimensional method accounting for misalignment - the Reynolds equation must be solved explicitly. To keep the simulation effort in a reasonable range, you must decouple the hydrodynamic solution from the dynamic solution of ADAMS/Solver. During the dynamic solution, ADAMS/Solver [9],subroutines do the database access and the necessary analytical steps (coordinate transformations, and so on) [1].

The component outputs give the following results: forces, velocities and displacements.The creation of a nonlinear FE model of roller bearing is a first step; however, MBS using the simpler model has to be created.

4. Interaction between reduced models

The solution of the complex model in time domain runs in ADAMS[9]. Major parts, for example a crankshaft and an engine block, are solved as flexible parts [4]. The other parts are solved as rigid. Gas forces on pistons and a torsional damper are added in ADAMS as well. Some temperature elements for dependency on temperature are defined. The first temperature element defines the dependency of the oil temperature in bearings on the engine speed, and is referenced by a viscosity element that describes the dependency of oil viscosity on oil temperature. The oil viscosity is then necessary to describe the behaviour of hydrodynamic bearings. The second temperature element defines the dependency of the torsional vibration damper temperature on engine speed, which is referenced again by the torsional damper.The solution is very complicated and requires a great deal of computer time[4]. The numerical solution of the equation of the motion in time domain runs with integration time step corresponding to a crank angular displacement of approximately 0.01°.

5. Diesel in-line 6-cylinder engine cranktrain

Accelerations, velocities and displacements of all parts, both rigid and flexible, can be obtained from the ADAMS solution. Length or torsional deformations and stresses of flexible parts can also be obtained in ADAMS. One of the major problems is the torsional vibration of the crankshaft [8]. Tab. 1presents critical engine speeds it can be determined from the first torsional natural frequency.The comparison of calculations and measurements is still a decisive quality criterionin the CAD environment. Non-contact measurements with laser Doppler vibrationmeters make it possible to verify complex calculation models. In-line 6-cylinder measurement was performed on an experimental engine.The use of viscous torsional vibration damper significantly reduces the vibration and the stress level. Rotational accuracy of valvetrain is the next problem [5]. The reduction of torsional vibrations decrease rotational accuracy and positively affects the combustion process[8].The viscoustorsional vibration damper decreases axial vibration of the crankshaft pulley.An acceleration sensor was used for crankcase sump vertical and longitudinal vibration measurement. The reduction of the vertical acceleration of the in-line 6-cylinder engine crankcase soup centre is more than 90 percent at nominal engine speed [6].The solution of reduction the complex model in time and frequency domain runs in ADAMS. Major parts are solved as flexible parts; see Figure 2[2], [7].

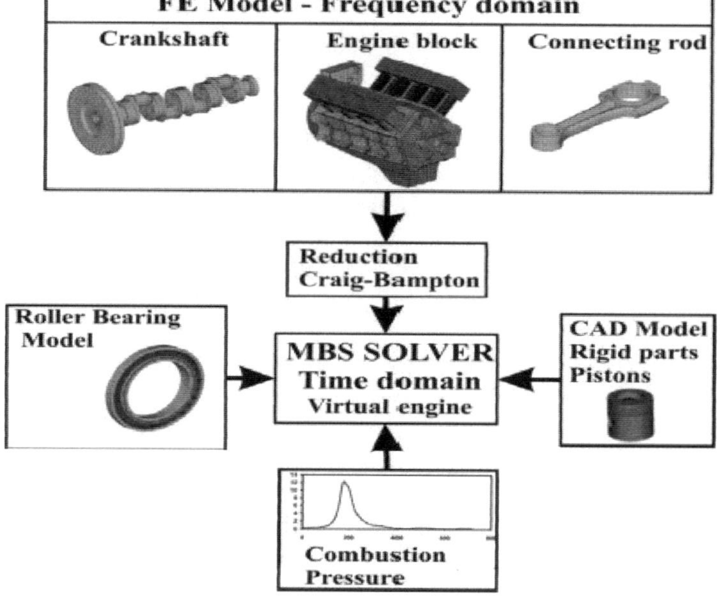

Figure 2. Assemble of a virtual engine with flexible and rigid parts

The torsional vibration order analysis of the pulley of the Diesel in-line engine with and withouta damper is presented in Figure3 [5], [7].

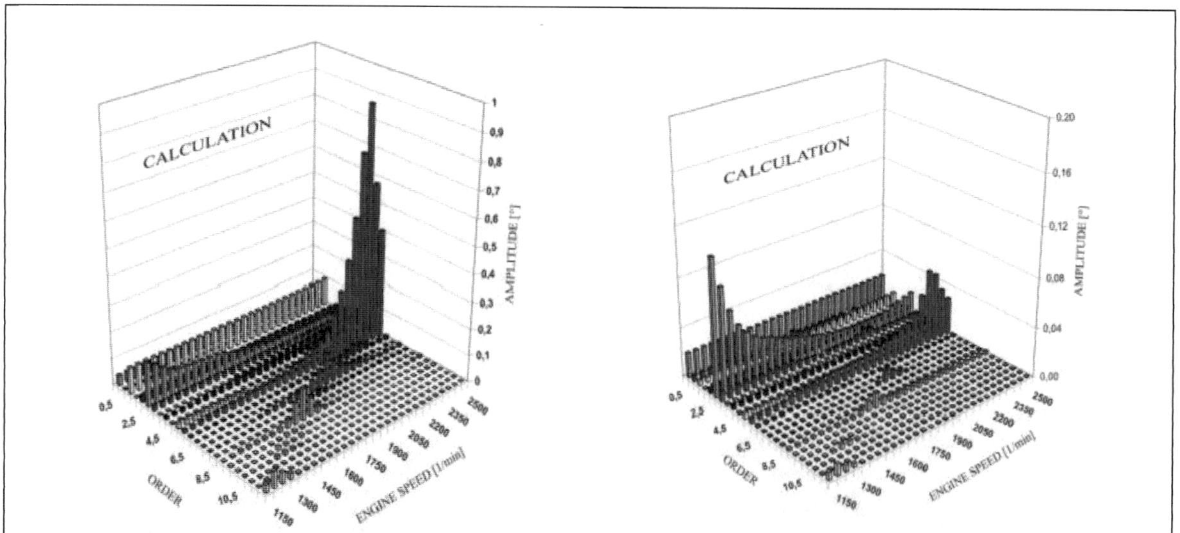

a) Pulley without damper	*b) Pulley with viscous torsional vibration damper*

Figure 3. Assemble Computed torsional vibration order analysis of Diesel in-line 6-cylinder engine

Table1. Critical engine speeds.

Order κ [-]	Computation n_{KRIT-C} [min^{-1}]	Measurement n_{KRIT-M} [min^{-1}]
3	4800	4750
6	2400	2370
6.5	2215	2188
7.5	1920	1896
9	1600	1580
12	1200	1185

6. Conclusion

Internal combustion engine simulation is a very complex problem consisting of many partial issues. The simulation of cranktrain dynamics is a central module of the virtual engine. New calculation methods are necessary to achieve more accurate results that allow a design that is nearer to the mechanical limit. The combination of ANSYS and ADAMS [9] provides a powerful implement for virtual engine design[5], [7]. With the increasing power of modern computers the new techniques will replace the traditional methods. Applied viscous torsional vibration dampers with extremely high viscosity silicone fluids are effective means of modern engine vibration and noise reduction[6]. Computational models of the Diesel in-line 6-cylinder engine were developed and verified by measurement on existing prototypes of engines and help to increase truck engine reliability. The paper contains partial methodology which was used in the construction of engine which has been tested and is ready for production. We expect its use as a propulsion unit of special vehicles.

Acknowledgments
Presented work has been prepared with support of the project MOBAUT, University of Defence Brno, Czech Republic

References

[1] SHABANA, A.*Dynamic of Multibody System.* Second Edition. Cambridge University Press. Cambridge, UK, 1998. ISBN 0 521 594464

[2] NOVOTNY, P., PISTEK, V., STODOLA, J. *Virtual Engine Tool for Ground Military Vehicle Reliability Increase.* Advances in Military Technology. Vol. 1. No. 1, 2006 (p 49 - 69)

[3] SHIGLEY, J, E., MISCHKE, CH. R., BUDYNAS,R.G. *Machanical Engineering Design.* New York. McGraw Hill Companies, 2004

[4] PISTEK, V.,NOVOTNY, P., STODOLA, J. *Virtual Engine Tool for Ground Military Vehicle Reliability Increase.* 10th European Armoured Fighting Vehicle Symposium. Shrivenham. UK, 2005

[5] STODOLA, J., STODOLA, P. *Virtual Design of Ground Armored Vehicle.* International Conference Transport Means 2009. Kaunas Lithuania. ISSN 1822–296 X (p 43 – 46)

[6] FURUHAMA, S., TAKIGUCHI, M. *Measurement of Piston Friction Force in Actual Operating Diesel Engine.* SAE Technical Paper 790855, 1979. Doi: 10.4271/790855

[7] NOVOTNY, P., BREZNICKA,A., STODOLA, J. *Virtual Engine - Tool for Military Truck Reliability Increase.* International conference Armament and Technics of Land Forces 2016. Liptovsky Mikulas, 2016. IBSN 978-80-8040-537-3 (pp 169 – 180)

[8] ROSS, D. Vibrations and Noise due to Piston-Slap in Reciprocating Machinery. Sound and Vibration., 2, 1965 (pp1 32 - 146)ADAMS [online]. [cit. 2016-11-7]
Available from>http://www.mscsoftware.com/product/adams

[9] ADAMS [online]. [cit. 2016-11-7]
Available from>http://www.mscsoftware.com/product/adams

IOP Conf. Series: Materials Science and Engineering **212** (2017) 012003 doi:10.1088/1757-899X/212/1/012003

Investigation of contact pressure in auto panel drawing dies during the forming process by using FEM method

FatihKoşar, Kemal Yaman

Depth. Of Mechanical Systems, Research and Development Institute, Turkish Scientific and Technical Research Council, (TÜBİTAK-SAGE), 06261, Ankara, Turkiye

E-mail: fatih.kosar@tubitak.gov.tr; kemal.yaman@tubitak.gov.tr

Abstract. In automotive stamping dies, the die design is generally based on experience. Usually, the loads on the die is not considered too much when the ribs, which strengthen the structure of the die, are positioned. Despite an abundance of manuscripts in the literature on formability and springback, the number of studies that examine the panel-die interaction is rather limited. The impact of the loads occurs in the form of contact pressure (CP) on the upper and lower die when the sheet metal is formed in a drawing die. For die designs, the CPs on the die must be calculated accurately in order to determine the correct position for the ribs and to optimize the weight of the die. The present study compares the CP distribution on a auto panel drawing die through different solution methods to examine the time-dependent change of the CP.

1. Introduction

Despite there are many studies about panel quality in the forming process, the number of studies on the metal-die interaction is somewhat sparse. In a die design, the positioning of the ribs is based on the experience and foresight of the designer. Although the weight of the vehicles has gradually decreased, no changes have occurred in the weights of the dies. The sole reason for this is because companies and academia focus more on panel quality, while die-panel interaction is not sufficiently investigated. However, the weight of stamping dies, as high as 40 tons, can be significantly reduced through optimization. This will allow cheaper dies and a larger variety of models to be offered to customers.
The interaction between metal and the die occurs in form of contact pressure (CP). In order to ascertain the stress occurring in the die during the forming process, it is first necessary to correctly determine these CPs. Nowadays, finite elements software can be used to easily determine the CPs.The objective of the present study is to determine the loads affecting a draw die in the form of CP and to examine the stresses created by such loads on the die. By inspecting the stress values on the die, conventional die design methodology can be questioned whether over-design or not. Due to low strain rates, the forming process is accepted to be quasi-static [1-5]; however, during forming process the CP values could change drastically. In the present study, the CP obtained through the quasi-static solution is compared with those obtained through the time-dependent solution.

There are some studies in literature about the effect of CP on die wear and tool life. Pereira et al. investigate the distribution of CP over the die radius, throughout the duration of channel forming

process for estimation of tool wear and life. Their study reveals that the CP response can be divided into two distinct phases that are steady state and transient phases. In steady state phase, CP remains almost constant however, in transient phase CP is more than double the magnitude of the steady state peak pressure. They also compared the quasi-static and transient FE simulation results for maximum CP distribution on die radius during the both phases. Quasi-static simulation results showed good correlation with the transient simulation results for both the transient and steady state regions [6]. In another study, Pereira et al. directly examines the contact sliding distance experienced during a typical sheet metal stamping process. Analysis of the contact conditions experienced by the blank showed that there is a region on the blank surface that experiences longer sliding distances at very high contact pressures. The analysis of the conditions at the die and blank surfaces both showed that the identified transient stage is likely to be critical to the overall wear/galling behavior [7-8].
A new slider type of test system was developed by Cora et al. [9] in order to replicate the actual stamping conditions including the CP state, sliding velocity level and continuous and fresh contact pairs (blank-die surfaces). Test results showed that the TD and CVD coatings have performed better wear resistance performance on contact surfaces.

2. Numerical Study

In this study, two different finite elements software were used to determine and compare the CPs. One of the software used is Autoform®, a program that specializes in metal forming, that is frequently used in the automotive sector, whereas the other is Abaqus®, which has strong modelling and solutions for non-linear analyses. The implicit method is used for the Autoform solution method and the explicit method is preferred for the Abaqus solution; thus, the outcomes of the quasi-static and time-dependent analyses of the CP were compared. In die analyzes the ANSA® software, which has the ability to create high convergence mesh network is used and for post processing the MetaPost® is used.

2.1. Form analysis

The forming analysis model of Abaqus explicit is shown in Figure 1 and Figure 2. All die elements, except the sheet metal, were accepted as rigid. The 3D CAD data of the die in the form analysis were obtained from a design, which were manufactured before.

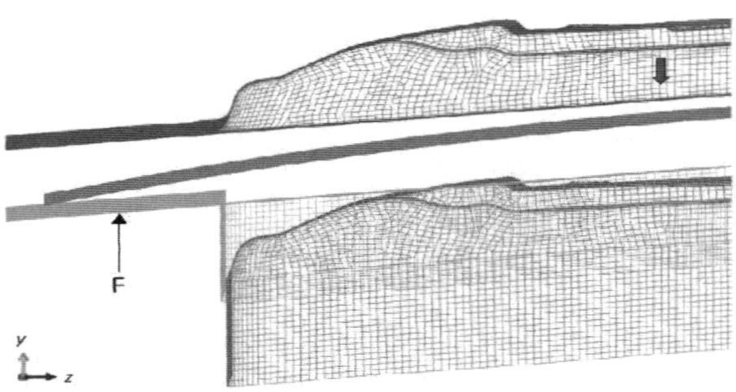

Figure 1. Form analysis model

The lightening holes on the upper die, which can be seen in the Figure 2 also included to forming analysis.

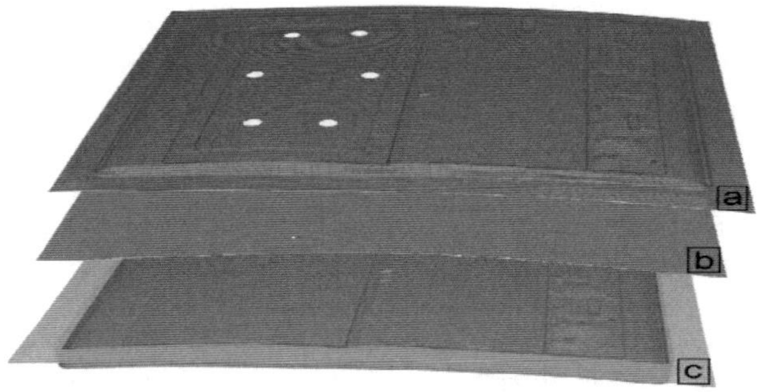

Figure 2. (a) Upper die, (b) sheet metal, (c) lower die and blank holder surfaces

The analysis model was built to allow for the sheet metal thickness (0.65 mm) when the model is in full-closed status. (See Fig. 2)

2.2. Model parameters

Some of the parameters and their properties that directly affect the analysis results are listed below:
Blank (Panel) Material: Material tensile test were performed for true stress-strain relationship of blank material. The Figure 3 shows the plastic hardening curve of blank material (bake hardenable steel CR240B2) obtained from the test.

Table 1. Properties of die material [10-11]

Die Material	Yield Stress [MPa]	Ultimate Stress [MPa]	Modulus of Elasticity [GPa]	Poisson ratio
EN-JS 2070	420	700	170	0.26

Die Material: Die material used in the analysis is EN-JS 2070 (GGG 70L) cast iron. The properties of die material are given in Table 1.
Total Time: Since the critical time increment of explicit analyses is very small, the total solution time may be longer when compared with the implicit methods. In forming analysis, total analysis time can be shortened by increasing the die speed far above the actual conditions, but artificial dynamic effects that cannot occur in the real world can occur. In metal forming analyses, a parameter is used to avoid such effects. These parameters are the explicit time step required for 1 mm movement of the die (i.e. ncpm: number of cycles per millimeter). A value between 100-1000 is recommended [12]. For the present study, this value was set to 100. Also, the maximum die speed recommended for explicit analyses is 2 mm/ms.
Speed: For the analyses, the upper die speed was defined as linear with first 2 ms and last 2 ms so that the first and the last speeds are both 0, hence reducing the dynamic factors.

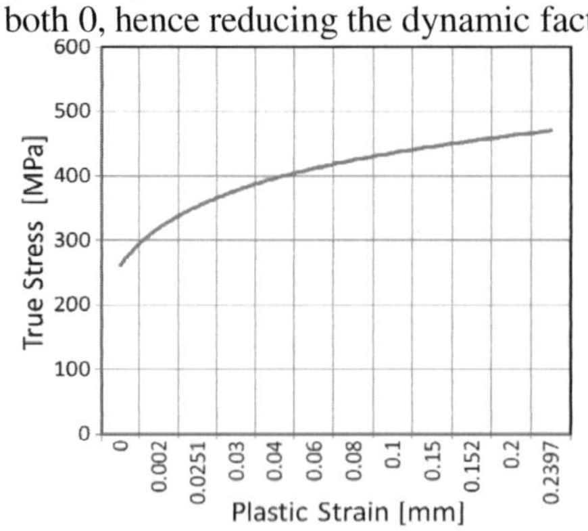

Figure 3. True stress-strain curve of blank material

IOP Conf. Series: Materials Science and Engineering **212** (2017) 012003 doi:10.1088/1757-899X/212/1/012003

Mesh: Since the element size needs to reflect the actual geometry and forms, an average 4 mm was chosen for lower die, upper die and blank holder and 3 mm was chosen for the sheet metal. S4R (shell quadrilateral reduced integration) was the element type selected for the metal sheet.
Contact Characteristics: For contact definition, isotropic penalty friction formulation was used. 0.15 friction coefficient was used in simulations. For contact stiffness, a "hard contact" pressure over-closure relationship is used.
Boundary Conditions: Lower Die rigid body reference point is fixed and a speed profile was assigned to the upper die.
Force: For the blank holder pressure force, which is one of the most important parameters of the form analysis, a force value of 2300 kN was used as provided by the Autoform. The gravitational acceleration was also included in the analysis.

3. Result and discussion

Figure 4 shows the maximum CPs obtained by scanning the time intervals in the Abaqus Explicit analysis and the maximum CP gained from Autoform solution for sample forms.

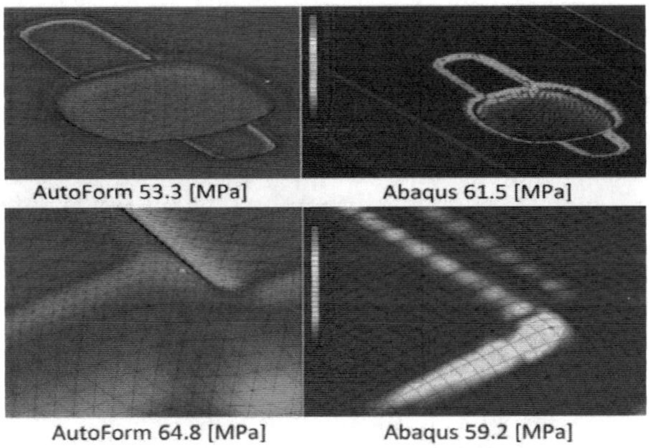

Figure 4.Abaqus/Autoform contact pressures comparison

When the Autoform and Abaqus CPs are compared, it is seen that zero CP occurs in flat sections of panel in both analyses. At local forms a difference of up to %15 are seen between the two results. The solution method (implicit/explicit), algorithms and mesh differences are the main reasons for the difference. CP change during the forming process for a "S" shape form can be seen in Fig. 5.

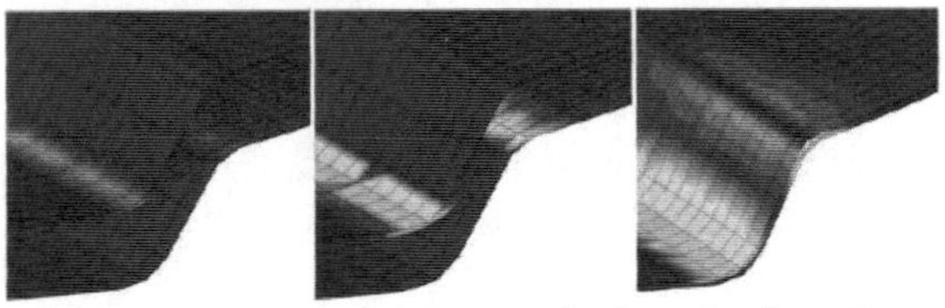

Figure 5. Contact pressure change during the forming process

Points labeled on the die surface as seen in Figure 6a. The time-dependent CP change of the points on this form are given graphically in Figure 6b.

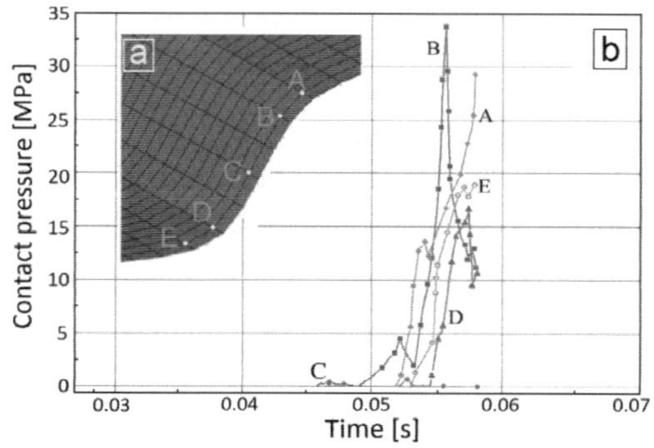

Figure 6. The time-dependent pressure change of the points (a) labeled on the die surface and (b) graphical results

Although high CPs are seen locally in the analysis of Autoform and Abaqus, the average value of both outcomes is concentrated around 60 [MPa]. In order to investigate the effect of CP on the lower die, a static analysis was performed using the maximum CP values obtained from the previous Autoform results. The panel CP values shown in Fig. 7a were applied to the relevant area on the lower die.

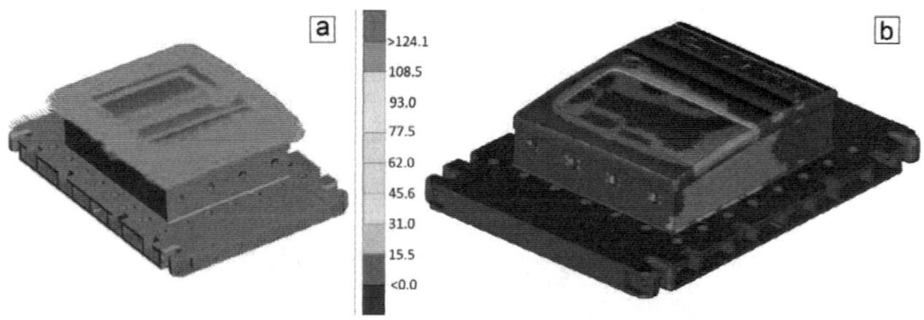

Figure 7. Boundary conditions of lower die for CP analysis (a) and CP analysis result on lower die (b)

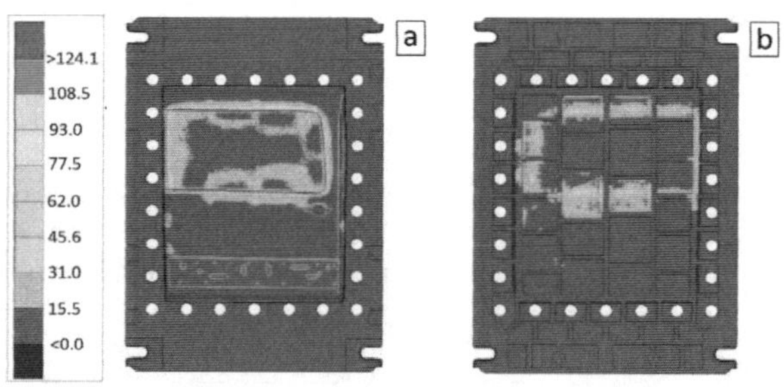

Figure 8. Result of CP analysis on lower die (a) top view (b) bottom view

The results of static analysis on lower die is shown in Fig.7b Stress distribution over the die geometry is also given in top and bottom view in Fig. 8a and Fig. 8b respectively. The highest stress values were formed on the surface of glass zone in parallel with the CP form analysis values.

4. Conclusion

In the present study, the CP obtained from Autoform Implicit and Abaqus Explicit solution method are compared. This comparison showed that the time-dependent and quasi-static CPs were almost similar. It was seen that the similar CP results obtained through different solution methods can be used for die-blank (panel) interaction. The stress results due to the obtained maximum CPs can be used for die fatigue analyses and die optimization studies. Furthermore, when the CP results are examined, it is seen that the forming process occurs through the male and female forms in the lower and upper die.

With this approach, the flat sections need not be pressed between the lower and the upper dies. In fact, in the Abaqus form analysis, the lightening holes located in the upper flat area were added to the calculations and there was no adverse effect on the panel quality of these sections. If the flat sections are removed on the lower and upper die, savings can be made in terms of material, machining and the heat treatment. As a result, for die design, the CPs on the die must be calculated accurately in order to determine the correct position for the ribs and to optimize the weight of the die.

Acknowledgments

The authors wish to thank to the FORD Otomativ San. A.Ş. Tool&Die Depth.for their motivations, supports of equipment and workforce in order to carry out the study.

References

[1] ABAQUS Inc. ABAQUS Version 6.14 Documentation, Dassault Systems, Providence, RI 029009, 2004.
[2] N. Rebelo, J.C. Nagtegaal, H.D. Hibbit, "Finite element analysis of sheet forming process", Int. j. for Num. Methods in Eng., vol. 30, pp.1739-1758, 1990.
[3] S.C. Tang, "Trends on simulation of sheet metal forming process", SAE Transactions: J. of Mater. & Manuf. Vol. 109, pp. 523-530, 2000.
[4] A.E. Tekkaya, "State-of-the-art of simulation of sheet metal forming", J. of Mater. Proc. Tech., vol. 103, pp. 14-22, 2000.
[5] M. Tisza, "Numerical modelling and simulation in sheet metal forming" J. of Mater. Proc. Tech., vol. 151, pp. 58-62, 2004.
[6] M.P. Pereira, W. Yan, B.F. Rolfe, "Contact pressure evolution and its relation to wear in sheet metal forming" Wear, vol. 265, pp. 1687-1699, 2008.
[7] M.P. Pereira, W. Yan, B.F. Rolfe, "Sliding distance, contact pressure and wear in sheet metal stamping" Wear, vol. 268, pp. 1275-1284, 2010.
[8] M.P. Pereira, W. Yan, B.F. Rolfe, "Wear at the die radius in sheet metal stamping" Wear, vol. 274-275, pp. 355-367, 2012.
[9] Ö.N. Cora, K. Namiki, M. Koç, "Wear performance assessment of alternative stamping die materials utilizing a novel test system" Wear, vol. 267, pp. 1123-1129, 2009.
[10] W. Callister, "Fundamentals of Materials Science and Engineering", 5th John Wiley & Sons, Utah, 2000.
[11] K.M. Zwilsky, "Properties and Selection Irons Steels and Performance Alloys", Vol.1 2nd ed., ASM, U.S.A, 1990.
[12] B.N. Maker, X. Zhu, "Input parameter for metal forming simulation using Ls-Dyna", Liwermore Soft. Tech. Co., 2000.

IOP Conf. Series: Materials Science and Engineering **212** (2017) 012004 doi:10.1088/1757-899X/212/1/012004

Vibration Reduction Design of Equipment on the Low-floor Tramcar's Roof

Feng Zunwei*, He Binbin

CRRC Nanjing Puzhen Co., Ltd. Bogie Design Department, Nanjing 210031, China

* levn0516@126.com

Abstract. In order to suppress the influence of the excitation source of the roof equipment on the vibration of the vehicle body, a vibration reduction design scheme based on rigid and flexible coupling dynamic model is proposed. The finite element model is established, and the finite element analysis is used to calculate the car body mode. The dynamic model is established and the vibration mode is analyzed. The frequency range of the equipment is determined by the analysis result which is based on the vibration reduction theory. The rigid and flexible coupling dynamics model is established to optimize the equipment support frequency. The results show that when the frequency of the equipment is 10 Hz, the support point can obviously eliminate the vibration effect. The dynamic deflection of the equipment is less than 2 mm, and the vibration performance satisfies the design requirements.

1. Introduction

Low floor tramcar belongs in transit traffic mode, with fast, convenient, safe advantages, which also save energy and investment, protect the environment. So low floor tramcar is very suitable for arterial traffic of small and medium-sized cities and artery traffic of large cities [1]. In this paper, the tramcar adopts a hybrid drive mode of super capacitor and fuel cell. The capacitor is omitted with this power supply method, which reduces the occupation of the urban space. But the fuel cell and heat sink need to be installed to the roof [2]. the equipment will influent the vibration of body [3-7]. At the same time, the equipment is excitation source. In this paper, a rigid-flexible coupling dynamics model is established to optimize frequency of the device.

2. Design considerations

The low floor tramcars are composed of three body modules. The center of each module is equipped with a bogie. The middle body module is the trailer, and the two side is motor car, which are group into Mc+TP+ Mc. The fuel cell and heat sink need to be installed to the roof of the middle car, which are shown in Figue.1. There are two pieces of fuel cell and a heat sink. A total of 12 installation points are built as shown in the figure1. The working frequency of the heat sink is shown in the table.1.

Table1. The working frequency of the heat sink.

Seasons	Commonly	Summer	Spring	Autumn
Working Frequency(Hz)	50	40	30	25

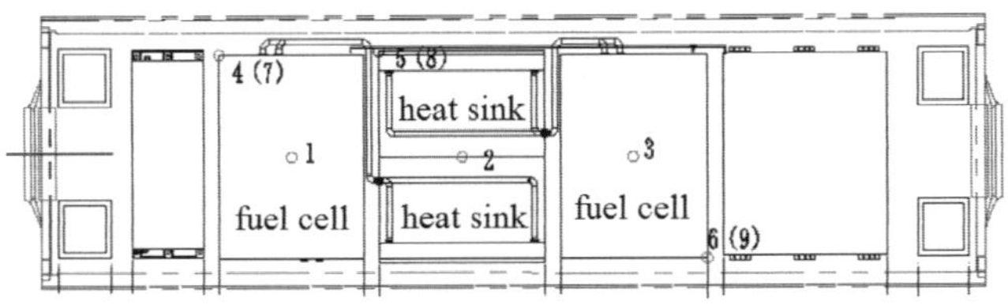

Figure 1. The fuel cell and heat sink of the roof.

In order to study the influence of the roof equipment, the model of three-module body is established. First, calculate the rigid body mode. Then establish the finite element model of the middle car. According to polycondensation calculation, a rigid-flexible coupling dynamics model is established. Then, study the influence of the roof equipment.

3. A rigid-flexible coupling dynamics model of car's body

In order to establish a rigid-flexible coupling dynamics model, the finite element model need to polycondensation calculation to improve the calculation efficiency. The polycondensation calculation follows the Guyana rule[8]. The points of polycondensation calculation are selected as figure.2.

Figure2. The selected points of polycondensation calculation.

After the polycondensation calculation, the top nine lower-order modes are shown as the table2.the results have errors up to 3.58%, which explain the high accuracy of the polycondensation calculation. It is up to the standard of the establishment of dynamic model.

Table 2. The main lower-order modes of the car body

Oder	The finite element model /Hz	Polycondensation calculation /Hz	error/%
1	24.22	24.12	0.41
2	26.76	26.14	2.32
3	28.84	28.71	0.45
4	31.87	33.01	3.58
5	33.64	33.45	0.56
6	44.21	45.20	2.24
7	45.84	45.72	0.26
8	47.09	47.99	1.91
9	48.81	48.74	0.14

A fixed hinge and a hinge are set between car B and car C, which makes the body B, C in the vertical is rigid, only rotate around the z axis. Car A, B could rotate around the y axis.Based on the coordinate system, the SIMPACK model is established. According to the topology of the vehicle system, the hinge between the car bodies can be simulated by spring. The SIMPACK model is replaced with the car body of polycondensation calculation. Then the rigid-flexible coupling model is obtained as the figure.3.

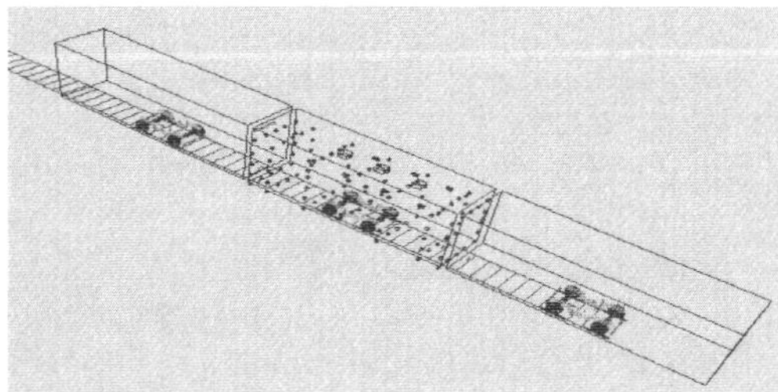

Figure3. Simpack multi-body dynamics model

In SIMPACK rigid body model, the body is linear. The vibration modes can be obtained as shown in the table3, there are six vibration mode of rigid body. The frequency of floating is 1.8988 Hz.

Table 3. The vibration modes

Oder	Shape	Frequency/Hz	Damping ratio /%
1	The rolling swing	0.452 8	9.64
2	Shaking of car A	0.472 2	63.48
3	Nod of car A	0.852 1	27.96
4	Traverse of car A & C	1.363 0	23.55
5	Floating	1.898 8	38.67
6	Floating of car A	1.928 1	29.78
7	Shaking of car A & C	2.058 0	9.48
8	The rolling swing	2.184 1	61.90

4. Vibration design theory and equipment frequency selection

According to the basic theory of vibration absorption designing, no matter how much damping ratio$\frac{C}{C_0}$, only when$\frac{\omega}{\omega_n} > \sqrt{2}$, the displacement transfer rate is less than 1. Therefore, in order to achieve the vibration isolation, the natural frequency of elastic support system $\frac{\omega}{\omega_n} > \sqrt{2}$ (condition 1); if the condition cannot be achieved, the natural frequency of elastic support system $\frac{\omega}{\omega_n} < 0.4$[9,10](condition 2).

The body vibration is divided into rigid vibration and elastic vibration. According to the analysis, the rigid mode, the rigid body vibration gathers in 0~3 Hz. The lowest vertical frequency is 2.184 1 Hz. The overall vibration frequency is 2.1841 Hz. The body elastic vibration is above 24.22Hz without the roof equipment. There isn't body partial model under 24.22Hz.

According to condition 1, $\omega_n < \frac{\omega_1}{\sqrt{2}}$, the first-order respiratory vibration frequency ω_1 is selected, $\omega_1 = 24.22$ Hz, then $\omega_n < \frac{\omega_1}{\sqrt{2}} = 17.13$ Hz.

According to condition 2,$\omega_n > \frac{\omega_2}{0.4}$, the whole floatation vibration frequency ω_2 is selected, $\omega_2 =$ 2.184 1 Hz, then $\omega_n > \frac{\omega_2}{0.4} = 5.46$ Hz.

For passive isolation, it is better that the vibration isolation frequency is between 6 and 17Hz.

5. Optimization of device frequency

First, the Sperling indicator of intermediate car is selected as the basic evaluation index. For evaluating the vibration of the device, the acceleration RMS of three points is selected. As figure 1, number 1~3 is the center of the front fuel cell, radiator, and the final fuel cell. Number 1 point is selected to calculate transmissibility. Number 4~6 points are selected to calculate displacement. Applying three-way sine excitation achieves the influence of steady work of radiator to train. Each direction of excitation amplitude is set 200N.

As shown in figure.4, there is different lateral, vertical Sperling indicators under the different work of radiator. The result shows that the lateral, vertical Sperling indicators increase with the frequency increasing. When the device frequency is 8Hz, the Sperling index obtains the optimal value.

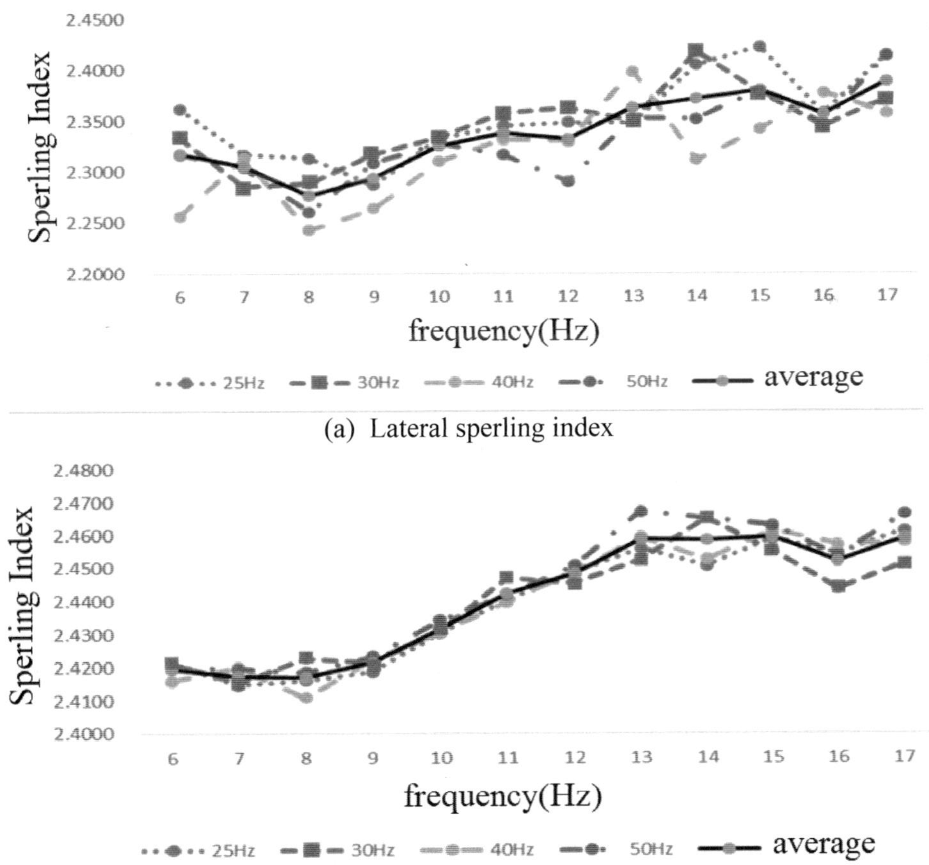

(a) Lateral sperling index

(b) Vertical sperling index

Figure 4. Sperling index of car body

With the frequency changing, the 1~3 point's lateral acceleration RMS (root mean square) value is shown as figure.5. The lateral acceleration increases with the equipment frequency increasing under 11 Hz for point 1 to point 3. The lateral acceleration decreases with the equipment frequency increasing after 11Hz for point 1 and point 3. When the equipment frequency is 6 Hz, the lateral acceleration is the lowest. But The lateral acceleration increases with the equipment frequency increasing for point 2.

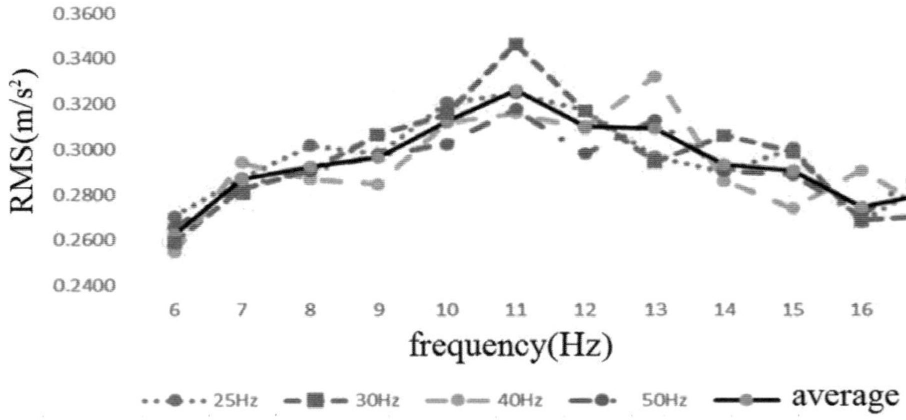

(a) The lateral acceleration rms value of point 1

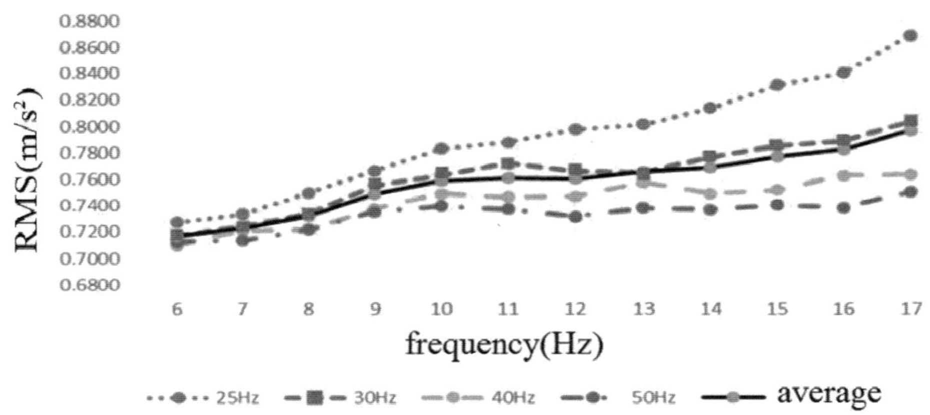

(b) The lateral acceleration rms value of point 2

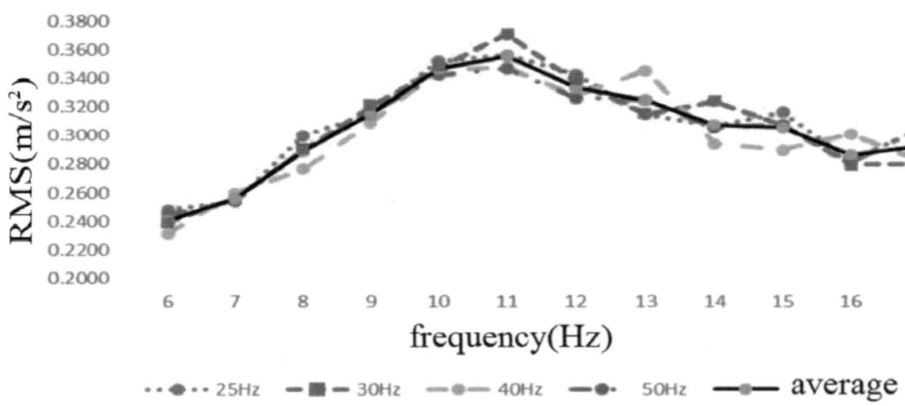

(c) The lateral acceleration rms value of point 3

Figure 5. The lateral acceleration rms value of point 1~3

With the frequency changing, the 1~3 point's vertical acceleration is shown as figure.6. The vertical acceleration increases, then decreases with the equipment frequency increasing under 11Hz for points 1~ 3. When the equipment frequency is 17 Hz, the vertical acceleration is the optimal value. But the vertical acceleration increases with the equipment frequency increasing for point 2.

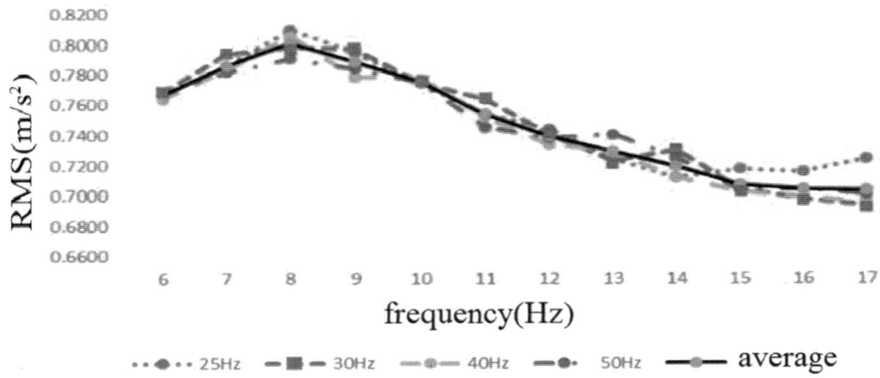

(a) The vertical acceleration rms value of point 1

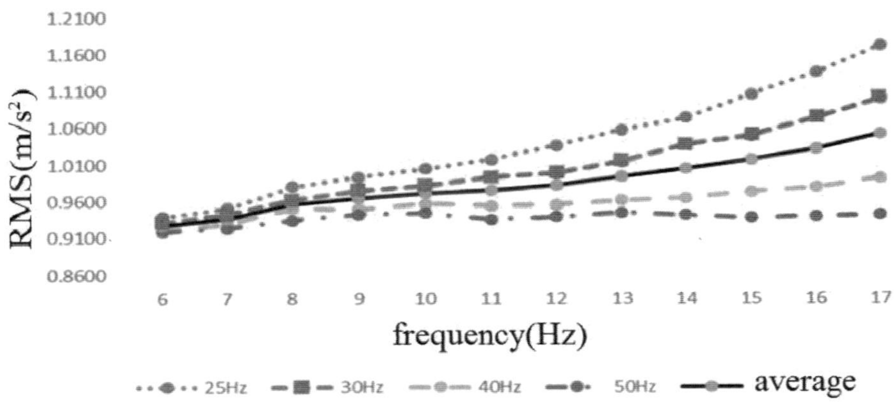

(b) The vertical acceleration rms value of point 2

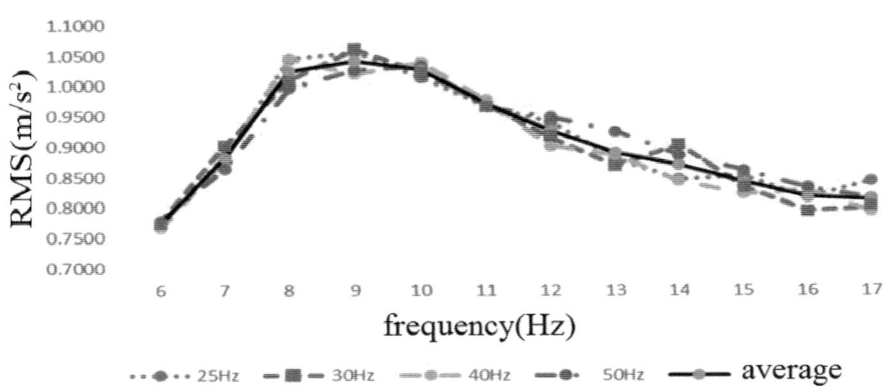

(c) The vertical acceleration rms value of point 3

Figure 6. The vertical acceleration rms value of point 1~3

IOP Conf. Series: Materials Science and Engineering **212** (2017) 012004 doi:10.1088/1757-899X/212/1/012004

6. Conclusions

The paper introduces a vibration reduction design method of low-floor tramcar equipment., which adopts a rigid-flexible coupling dynamics model by SIMPACK, the equipment frequency is calculated. The frequency of the equipment is optimized by the calculation of the stationary and dynamic deflection. This paper provides a theoretical reduction for low floor tramcar's roof equipment.

References

[1] Minghua Zhao, Xiping Niu, Danyan Yang, Chunyou Gao. Domestic Independent-developed 100% Low Floor Light Rail Vehicle. Electric Drive For Locomotives,2013 (3): 59-63.

[2] Junjie Shi, Ming Li, Nan Shao. Analysis on Dynamic Performance of Hybrid Power System of Complex Railway. Railway Locomotive & Car,2014,34(3): 57-61.

[3] YU Jian-yong, ZHANG Li-min, HUANG Xiao-yu, TANG Qin. The Influence of the Hanging Device on Complete Car's Model. Noise and vibration control,2012,32(5): 97-100.

[4] Deng Hai,Gong Dao, Zhou Jinsong,Ma Minna. On Vibration Isolation Design and Test of Suspended Equipment on High-speed Railway Vehicle. Research on Urban Rail Transit,2015,18(2): 44-48.

[5] YANG Wen-fang WEI Qiang ZHU Lan-qin. Anti-Vibration design for an airborne electronic equipment based on finite element method. JOURNAL OF VIBRATION AND SHOCK,2010,29(5): 230-234.

[6] GONG Dao ZHOU Jin-song SUN Wen-jing GU Yue-jia. Impacts of hanging equipments on vertical riding stability of elastic high-speed train bodies. Chinese Journal of Construction Machinery,2011,9(4):404-409.

[7] ZHANG Xiang-ning LI Min-gao. Study on Rubber Isolator of Hoisting Device for Hanging EMU Equipment. Journal of Dalian Jiaotong University,2012,33(5):19-22.

[8] GUYAN R J . Reduction of Stiffness and Mass Matrices. AIAA Journal ,1965.

[9] Gong, D., Zhou, J., & Sun, W. (2016). Influence of under-chassis-suspended equipment on high-speed emu trains and the design of suspension parameters. Proceedings of the Institution of Mechanical Engineers Part F Journal of Rail & Rapid Transit, 230(8).

[10] Gong, D., Zhou, J. S., & Sun, W. J. (2013). On the resonant vibration of a flexible railway car body and its suppression with a dynamic vibration absorber. Journal of Vibration & Control, 19(5), 649-657.

IOP Conf. Series: Materials Science and Engineering **212** (2017) 012005 doi:10.1088/1757-899X/212/1/012005

Stiffness Parameter Design of Suspension Element of Under-Chassis-Equipment for A Rail Vehicle

Ma Menglin*, Wang Chengqiang, Deng Hai

CRRC Changchun Railway Vehicle Co., Ltd. National Engineering Research Center of Railway Vehicles, Changchun 130062, China

* mamenglin@cccar.com.cn

Abstract. According to the frequency configuration requirements of the vibration of railway under-chassis-equipment, the three- dimension stiffness of the suspension elements of under-chassis-equipment is designed based on the static principle and dynamics principle. The design results of the concrete engineering case show that, compared with the design method based on the static principle, the three- dimension stiffness of the suspension elements designed by the dynamic principle design method is more uniform. The frequency and decoupling degree analysis show that the calculation frequency of under-chassis-equipment under the two design methods is basically the same as the predetermined frequency. Compared with the design method based on the static principle, the design method based on the dynamic principle is adopted. The decoupling degree can be kept high, and the coupling vibration of the corresponding vibration mode can be reduced effectively, which can effectively reduce the fatigue damage of the key parts of the hanging element.

1. Introduction

At present, in the process of stiffness design of railway vehicles, the vibration decoupling between the under-chassis-equipment and the vehicle is mainly carried out. Zhang et al [1] found that the under-chassis-equipment operation under severe force, based on vehicle-track coupling dynamics model and mode calculation, select the suspension parameters of suspension elements, the stiffness and damping parameters of the suspended elements are determined according to the characteristics of the rubber and its installation space. Gong et al. [2-4] established a vertical rigid-flexible coupled dynamic model of the railway vehicle including the under-chassis-equipment, designed the parameters of the elements, and analyzed the parameters of the suspension elements based on vibration isolation theory, Running stability and the vibration of the under-chassis-equipment itself are analyzed for different equipment mass and suspension position on the vehicle. However, in the design of the under-chassis-equipment parameters of the railway vehicle, the under-chassis-equipment is regarded as the single-degree-of-freedom mass unit. This kind of design method does not take into account the coupling vibration of the under-chassis-equipment due to the equipment eccentricity. In fact, the under-chassis-equipment is generally a large box, or a box containing multiple equipment, so the eccentric state of the under-chassis-equipment is in general. Based on the actual conditions of the project, this paper considers the eccentric state of the under-chassis-equipment in detail, combining with the engineering case and design the three- dimension of the suspension element based on the static and dynamic principles respectively. The design method can be widely applied to the stiffness design of the under-chassis-equipment of railway vehicles.

2. Design method based on static principle

In the statics, when the under-chassis-equipment is in the eccentric state, still meet the horizontal and vertical torque reciprocal theorem. The eccentric load due to the eccentric state can be calculated reversely, and the stiffness of the suspension element is designed. Fig. 1 is a top view of a under-chassis-equipment and suspension elements, and a positive direction is defined as a bit having a small number and a L-side being a 1-bit side.

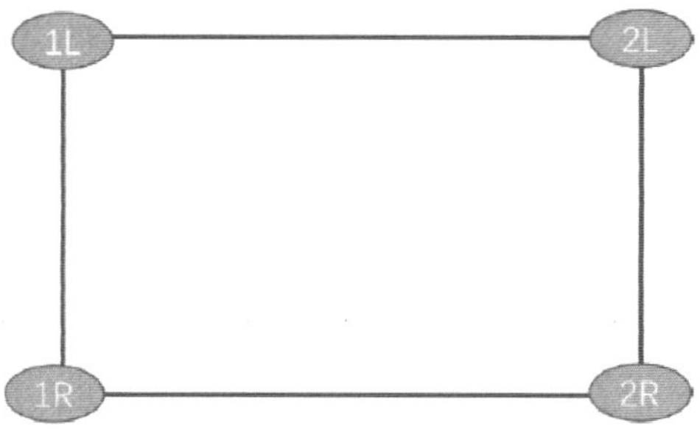

Figure 1. Under-chassis-equipment and its suspension elements of the top view

Based on the principle of static design method to design the three- dimension stiffness of the suspension elements, the general requirements with the following design parameters: Equipment total mass m_T, the number of suspension elements n, equipment vertical vibration natural frequency f_z, static and dynamic stiffness ratio R_z, suspension elements from the device centroid vertical distance x_{e1} and x_{e2} (Engineering parameters are generally in the form of longitudinal eccentricity and longitudinal span, can be converted to x_{e1} and x_{e2}), Lateral eccentricity y_e and lateral span y_L, Combined with the characteristics of the rubber element, the lateral stiffness is designed according to 1/3 of the vertical stiffness. Considering the general existence of the stopper device in the longitudinal direction of the under-chassis-equipment, the longitudinal stiffness is designed according to double of the vertical stiffness.

According to the stiffness design method of the traditional rubber element, the natural frequency of the vertical vibration of the equipment is related to the vertical dynamic and static stiffness of each suspension element as follows.

$$k_{dz} = m_T (2\pi f_z)^2 / n \tag{1}$$

$$k_{stz} = k_{dz} / R_z \tag{2}$$

The vertical average static load of each suspension component is:

$$F_{stz} = m_T * 9.81 / n \tag{3}$$

When the under-chassis-equipment is under the eccentric state, as a result of vertical eccentric and lateral eccentric load are:

$$F_{exz} = \frac{abs(abs(x_{e1}) - abs(x_{e2}))}{abs(x_{e1}) + abs(x_{e2})} * F_{stz} \tag{4}$$

$$F_{eyz} = \frac{abs(y_e)}{abs(y_L)} * F_{stz} * 2 \tag{5}$$

Accordingly, the suspension components of the under-chassis-equipment actually bear the vertical static load are:

$$\begin{cases} F_{1Lz} = F_{stz} + F_{exz} - F_{eyz} \\ F_{1Rz} = F_{stz} + F_{exz} + F_{eyz} \\ F_{2Lz} = F_{stz} - F_{exz} - F_{eyz} \\ F_{2Rz} = F_{stz} - F_{exz} + F_{eyz} \end{cases} \tag{6}$$

When the suspension element 1L to withstand the vertical static load is F_{1Lz}, the vertical dynamic stiffness is:

$$k_{dz1L} = F_{1Lz} / F_{stz} \cdot k_{dz} \tag{7}$$

In the same way, the vertical dynamic stiffness of suspension elements 1R, 2L, 2R satisfying the reciprocal theorem of moment can be obtained. The lateral and longitudinal dynamic stiffness k_{dz1R}, k_{dz2L}, k_{dz2R} of each suspension element are designed according to 1/3 and 2 times of its vertical stiffness respectively.

3. Design method based on dynamic principle

According to the physical model of the under-chassis-equipment studied in this paper, the dynamic model of the under-chassis-equipment can be established as shown in Fig.2. Considering the under-chassis-equipment as a rigid body with 6 degrees of freedom, the suspension element is simplified as a spring with three-dimensional stiffness and damping, set the mass center of the under-chassis-equipment as the coordinate origin, where the positive direction of x and y are defined as 1-bit terminal and 1-bit side.

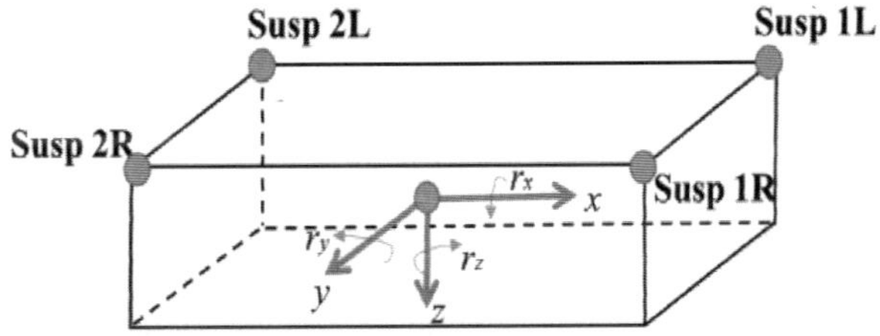

Figure 2. Dynamic model of the under-chassis-equipment

According to the dynamic model of the under-chassis-equipment, the sixth order mode vector and the modal frequency are solved.

$$\boldsymbol{KT}_j = \lambda_j \boldsymbol{MT}_j, \quad j = 1,2,3,4,5,6 \tag{8}$$

The j_{th} vibration energy distribution matrix of under-chassis-equipment is defined as [5-6]:

$$\boldsymbol{EM}(p,k) = f_j^2 \boldsymbol{M}(p,k) \boldsymbol{T}_j^p \boldsymbol{T}_j^k / 2 \tag{9}$$

The energy allocation of the j_{th} mode is:

$$\boldsymbol{EP}(k,j) = \frac{\sum_{p=1}^{6} \boldsymbol{EM}(p,k)}{\sum_{k=1}^{6} \sum_{p=1}^{6} \boldsymbol{EM}(p,k)} \tag{10}$$

The basic principle of the design method based on the dynamic principle is to make the vibration of the under-chassis-equipment independent of the design frequency. For any group of vectors $X=\begin{bmatrix}k_{x1} & k_{x2} & k_{x3} & k_{x4} & k_{y1} & k_{y2} & k_{y3} & k_{y4} & k_{z1} & k_{z2} & k_{z3} & k_{z4}\end{bmatrix}^{T}$ composed of the three-dimensional stiffness of the rubber elements, the decoupling degree and the modal frequency can be calculated

$$\begin{aligned} &\min f(X)=\sum_{j=1}^{6}\alpha_j(1-\max_{k=1,2,\ldots6} EP(k,j))^2+\sum_{j=1}^{6}\beta_j((f_j(X)-f_{jo})/f_{jo})^2 \\ &s.t.\ X^L\le X\le X^U \\ &\quad f_j^L\le f_j(X)\le f_j^U\ ,\qquad j=1,2,\ldots6 \end{aligned} \tag{11}$$

As genetic algorithm has the characteristics of strong searching ability and good ergodicity [7-8], this paper will use genetic algorithm to solve (11).

4. Case study

According to a factory to provide under-chassis-equipment parameters and frequency configuration requirements, the under-chassis-equipment is suspended by four suspension elements, vertical vibration design frequency is 8Hz, based on the principle of static design, the three-dimensional dynamic stiffness of the suspension elements is shown in Table 1. The lateral stiffness is designed according to 1/3 of the vertical stiffness, and the longitudinal stiffness is twice as high as the vertical stiffness.

Table 1. The three-dimension dynamic stiffness of the suspension elements designed according to static principle

Suspension element	three- dimension stiffness		
location	Kx(N/mm)	Ky (N/mm)	Kz (N/mm)
1L	1304	217	652
1R	1410	235	705
2L	1022	170	511
2R	1128	188	564

Compared with the design method of the static principle, the design method based on the dynamic principle supplements the three-direction rotational inertia of the under-chassis-equipment. The coordinate parameter of the suspension element changes from two-dimensional to three-dimension, and the stiffness ratio is to give a certain numerical range rather than specific values. Table 2 is the three-dimensional dynamic stiffness of the suspension elements designed based on the dynamic principle. From the numerical results, it can be seen that, compared with the design method based on the static principle, the three-way dynamic stiffness of the suspension elements designed by the dynamic principle is more uniform. The next section of this paper will analyze the coupling vibration of the under-chassis-equipment when the two design methods were taken.

Table 2. The three-dimension dynamic stiffness of the suspension elements designed according to dynamic principle

Suspension element	three- dimension stiffness		
location	Kx (N/m)	Ky (N/m)	Kz (N/m)
1L	5.419e5	2.877e5	6.706e5
1R	5.373e5	2.024e5	6.868e5
2L	5.417e5	2.099e5	5.737e5
2R	5.511e5	1.885e5	5.690e5

5. Coupling vibration evaluation of the under-chassis-equipment

According to the analysis of the decoupling degree, the results of Modal frequency and decoupling degree based on two principles are shown in Table 3 and Table 4, respectively. From Table 3 and Table 4 can be seen the float vibration mode frequencies were 8.00Hz and 8.09Hz, and calculation results are in accordance with the design frequency. From Table 3, it can be seen that the design method based on the principle of static does not fully consider the coupling vibration of each degree of freedom of the under-chassis-equipment. The decoupling degree of the second and fourth order modes is 68.8 %, Nodal and telescopic vibration are serious. From Table 4, it can be seen that the decoupling degree of the vibration modes of the second and fourth order increases by 86.6% when the design method based on the dynamic principle is adopted, and the nodal and telescopic vibration are effectively reduced, thereby effectively reducing the key parts of the suspension elements of fatigue damage.

Table 3. Modal frequencies and decoupling degree based on static principle

Order		**1**	**2**	**3**	**4**	**5**	**6**
Natural frequency/Hz		4.56	7.10	8.00	17.28	21.00	29.72
Distribution of the energy %	x	0.0	**68.8**	0.0	31.2	0.0	0.0
	y	**99.9**	0.0	0.0	0.0	0.1	0.0
	z	0.0	0.0	**100.0**	0.0	0.0	0.0
	α	0.1	0.0	0.0	0.0	**99.9**	0.0
	β	0.0	31.2	0.0	**68.8**	0.0	0.0
	γ	0.4	0.0	0.0	0.0	0.0	**100.0**

Table 4. Modal frequencies and decoupling degree based on dynamic principle

Order		**1**	**2**	**3**	**4**	**5**	**6**
Natural frequency/Hz		4.77	6.08	8.09	13.71	20.25	21.34
Distribution of the energy %	x	0.0	**86.6**	0.2	13.3	0.0	0.0
	y	**99.9**	0.0	0.0	0.0	0.0	0.1
	z	0.0	0.1	**99.8**	0.0	0.0	0.0
	α	0.1	0.0	0.0	0.0	0.0	**99.9**
	β	0.0	13.3	0.0	**86.7**	0.1	0.0
	γ	0.4	0.0	0.0	0.0	**99.9**	0.0

6. Conclusions

(1) Based on the principle of static mechanics and dynamics, according to the design requirements, respectively, to design the three-dimension stiffness of the suspension elements of the under-chassis-equipment. From the numerical results, the two design methods can be applied to specific engineering

cases, compared with the design method based on the static principle, the three-dimension Stiffness is more uniform when dynamic principle is adopted.

(2) Based on the dynamic equation of the under-chassis-equipment and the calculation of the vibration frequencies and the decoupling degree, the coupling vibration of the under-chassis-equipment of two design methods is evaluated. The results show that the vibration frequencies of the under-chassis-equipment are close to the design frequencies. Compared with the design method based on the static principle, the design method based on the dynamic principle is adopted. The coupling vibration of the corresponding vibration mode can be reduced effectively, and the fatigue damage of the key parts of the suspension elements can be effectively reduced.

7. References

[1] X Zhang, M LI. Study on rubber isolator of hoisting device for hanging EMU equipment. Journal of Dalian Jiaotong University, 2012, vol.5, pp. 19-22.

[2] D Gong, J Zhou, W Sun. Modes matching between suspended devices and car body for a high-speed railway vehicle. Journal of Vibration and Shock, 2014, vol.8, pp. 180-185.

[3] D Gong, J Zhou, W Sun. Influence of under-chassis-suspended equipment on high-speed EMU trains and the design of suspension parameters. Proceedings of the Institution of Mechanical Engineers Part F Journal of Rail & Rapid Transit, 2016, 230(8), pp. 1790-1802.

[4] D Gong, J Zhou, W Sun. Impacts of hanging equipments on vertical riding stability of elastic high-speed train bodies. Chinese Journal of Construction Machinery, 2011, vol.4, pp. 404-409.

[5] Ma F, Morzfeld M, Imam A. The Decoupling of Damped Linear Systems in Free or Forced Vibration[J]. Journal of Sound & Vibration, 2010, 329(15):3182-3202.

[6] Jeong T, Singh R. Analytical Methods of Decoupling the Automotive Engine Torque Roll Axis[J]. Journal of Sound & Vibration, 2000, 234(1):85-114.

[7] Holland J H. Adaptation in natural and artificial systems: An introductory analysis with applications to biology, control, and artificial intelligence[M]. Michigan Press,1975.

[8] Goldberg D E, Others. Genetic algorithms in search, optimization, and machine learning[M]. Addison- Wesley Reading Menlo Park, 1989.

IOP Conf. Series: Materials Science and Engineering **212** (2017) 012006 doi:10.1088/1757-899X/212/1/012006

Fabrication and Mechanical Properties of Porous CP-TiUsing Selective Laser Melting (SLM)

Yaling Wang[1], Chunyu Zhang[1]*, PengZhang[2], WeiFeng[3], Xianshuai Chen[1], Jiaming Huang[1]

[1]Guangzhou Janus Biotechnology Co., Ltd., Floor 1, Building C, No.1121.,Nansha. Guangzhou, 511400, China
[2]Foshan stomatologicalhospital, Foshan, China,528000,China

[3]Shenzhen Institutes of Advanced Technology, Shenzhen, 518055,China

yl.wang@gzjanus.com,xuefeng851231@qq.com

Abstract. Recently, porous structure has become focus for the study of implantable devices due to the superior ability of osseointegration. Meanwhile, Selective laser melting (SLM), a versatile rapid manufacturing method, is widely utilized in medical restoration field due to its flexibility. Given the information about mechanical properties and manufacturing process of porous structure is limited, the aim of this study is to investigate the manufacturing process and static mechanical properties. Therefore, regular hexahedron structures are designed and fabricated by SLM, also the dimensions and mechanical properties are evaluated by experiments. The results show that manufacturing process and design of structure have significant influence on porosity and mechanical properties. Porous structure would reduce the elastic modulus and yield strength, and the ductility of porous structure is poor than that of solid structure. Moreover, the suitable pore size is >0.7mm. In conclusion, porous implant has great potential to use in bone repair and further study should be done to obtain more information.

1. Introduction

In recent years, with the development of RP technology, selective laser melting (SLM) technique comes into being, which makes the use of high-energy density laser beam to melt the metal powder completely, layer upon layer increment, ends up with computer aided design model of the metal products[1]. SLM is an advanced AM technique that can fabricate dense parts by completely melting metal powder to produce implant with material properties closing to those of the bulk materials [2]. Fabrication by SLM can manufacture specimens with whatever geometric shapes thanks to the design freedom and high reproducibility [3]. Compared with other conventional manufacturing, SLM has irreplaceable advantages and become a hot research topic [4]. Nevertheless, the SLM study is still in its infancy, the mechanical properties, accuracy and materials are subject to further discussion.
Dental implants are widely used today as aviable and reliable alternative for tooth replacement in edentulous patients and have been drawing more and more attention due to their advantages of reliability and comfort[5-6]. With the wide-ranging application of dental implant, its initial stability has been considered to be the primary condition of the success by most of scholars [7].

However, the implants fabricated by metallic biomaterials are much stiffer than natural bone, resulting in stress shielding and even bone resorption. The bone resorption caused by stress shielding isconsidered to contribute to undesired effects. Therefore, finding optimal structures to avoid the appearance of stress shielding is necessary. Porous structures are suitable for implantation since they allow for bone ingrowth through the open porosities and improve stability with the high roughness and lower stiffness and thus avoid stress-shielding [8]. SLM using to fabricate porous structure has the advantages to combine the solid materials and porous materials in one structure.

Since the method of fabrication and design would influence the clinical effect. In this study, the effect of design, fabrication and mechanical properties are analyzed to evaluate the performance of porous structure.

2. Materials and methods

2.1. Materials

A large number of studies have demonstrated when doing small deformation and static analysis, materials used in study are assumed to be continuous, homogeneous and isotropic. Titanium is the material of choice due to its favorable mechanical and biocompatible properties, with the tensile strength and yield strength of 383MPa and 475MPa based on SLM process. CP Ti grade 2 is used in this study, chemical composition is listed in Table1, with the purity of 99.8%. In this study, spherical commercially pure grade 2 Ti powder with particle size ranging from40 μm to 60 μm is used.

Table1. Chemical composition of CP-Ti grade 2 used in the study.

Element	Ti	Fe	C	N	H
w%	Balance	0.03	0.007	0.015	0.0035

2.2. Design and fabrication

Based on reverse engineering and 3D modeling software-Solidworks 2014, experimental models are shown in Fig.1. Ti 250-750 has a theoretical strut size of 250μm and pore size of 750μm. Rectangular porous specimen with dimensions of 10.25mm*10.25mm*15mm and porosity of 77% are fabricated for accuracy and mechanical properties analysis.

Ti 250-750

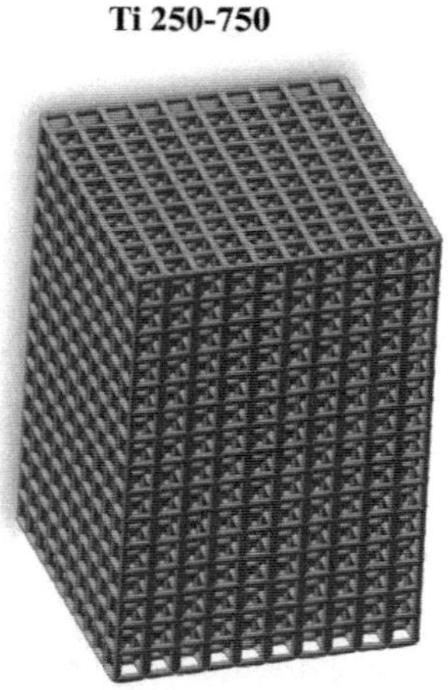

Figure 1. Additively manufactured porous CP Ti structures: 3D CAD visual representation

SLM125HL YLR-100-WC of SLM Solution is used to study SLM technique in this study. The sizes of the table are 125 mm×125 mm×75 mm, and the processing speed is $15cm^3/h$, layer thickness could be controlled from 20μm to 75μm, the scanning spot size is in the range of 70~130μm, the minimum processing thickness is from 140μm to 160μm. Based on the experimental results, the optimum technical parameters are shown in Table2.

Table2. Processing parameters of SLM technique.

process parameters	laser powder	scanning interval	scanning speed	layer thickness	Scanning spot
value	100w	0.13mm	255mm/s	70μm	87μm

2.3. *Mechanical testing*

Static mechanical testing for five rectangular specimens of porous structures is carried out based on the standard ISO 13314:2011. All tests are done using MTS Landmark370.10 by applying a constant deformation rate of 0.1mm/s (Fig.2).

3. Results

3.1. *Accuracy of manufacturing for dimensions and porosity*

Five specimens are fabricated by SLM and measurement results of dimensions by werth SCOPE-CHECK 300 composite optical measuring machine are shown in Table 3. The average dimension of pore is 0.66812mm and strut is 0.33262mm, with the average difference of 0.08mm. Archimedes measurements of five rectangular samples are used to evaluate actual porosity. Theoretical density of pure Ti is 4.5 g/cm^3 in this study. As a result, the actual porosity of pure Ti specimen is 72%, closing to the theoretical porosity of 77%.

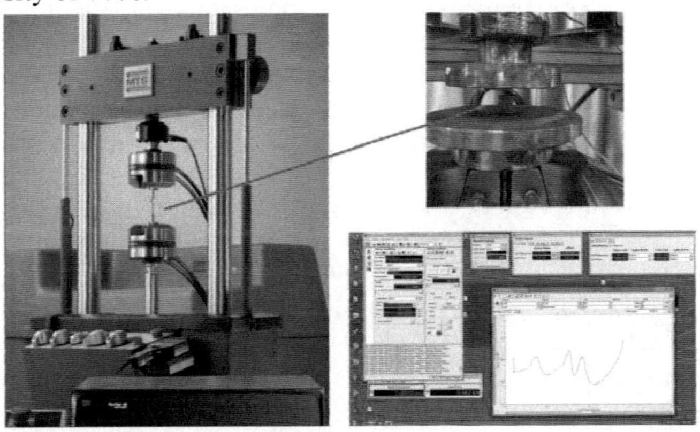

Figure 2. Machine and parameters of mechanical testing.

Table3. Dimensions of specimens by SLM (mm).

	Specimen-1	Specimen-2	Specimen-3	Specimen-4	Specimen-5
Pore	0.6652	0.6634	0.6723	0.6712	0.6685
Strut	0.3361	0.3354	0.3256	0.3317	0.3343

3.2. *Static mechanical testing*

The results of the static compression testing are summarized in Fig.3. A representative curve of strain-stress during static compression testing is shown to analyze the static mechanical properties. Different from the compression of solid structure, maximum compressive stress and strain could be registered due to the poor ductility of the porous CP Ti material. The maximum compressive stress is 186MPa, and σ_{pl}(the arithmetical mean of the stresses between 20% and 30% compressive strain)is87MPa.

Elastic gradient (the slope of linear deformation region within 20% and 70%plateau stress) is registered as 83.45GPa.In this context, the quasi-elastic gradient will not be referred to as stiffness. Eventually, σ_y(yield stress) is calculated as 170.3MPa based on the above data.

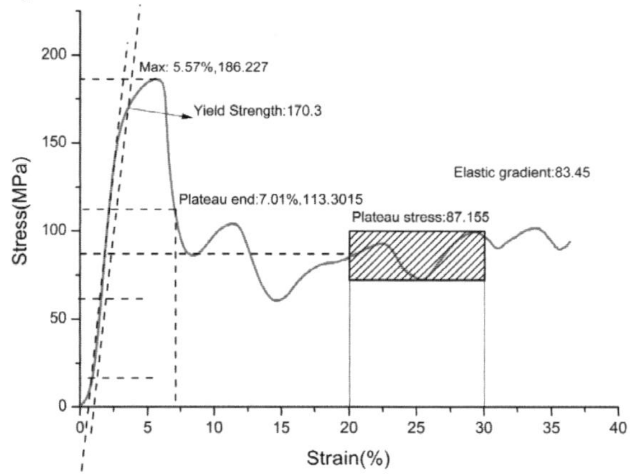

Figure 3. Strain-stress curve of static compression testing.

4. Discussion

4.1. Effect of design

Previous studies have illustrated that increasing the porosity would reduce stiffness of biomaterials. Highly porous biomaterials and large pore size provide ample space for bone ingrowth. Based on early studies, the minimum pore size is considered to be ~100um because of cell size, migration requirements and transport[9].However, acceleration studies suggest the pore size should be >300um. Small pores favored hypoxic conditions and induced osteochondral formation before osteogenesis, while large pores, that are well-vascularized, lead to direct osteogenesis (without preceding cartilage formation). Subsequent studies have shown researchers create scaffolds with pore sizes between 500 and 710 um to promote bone formation.Combined with the results by Ishaug-Riley SL, Bael SV, Otsuki B and Li JP, the proper size of pore is >0.7mm.

In static compression testing, it is observed that CP Ti porous structures continuously deform under increased compressive load, eventually reaching a first local maximum. This mechanical behavior is very similar to what is reported Ti6Al4V. The elastic gradient is not suitable for replacing elastic modulus of implant. Therefore, the formula (1) of Ashby and Gibson by previous study is used to recalculate the elastic modulus [10].

$$E^* = E * (1 - \varphi)^2 \quad (1)$$

Where $E^*$and E are the relative elastic modulus of porous structure and elastic modulus of metal, φ is the porosity. In this study, E^* is considered as 8.6GPa based on the porosity of 72% and the elastic of CP Ti 110GPa. The value of 8.6GPa is close to that of bone tissue (13.7GPa), resulting in a reduction of stress shielding.

4.2. Effect of fabrication on mechanical properties

In this study, CP Ti porous structures based on hexahedral unit cell with the porosity of 77% are fabricated to evaluate the processing precision and mechanical properties. In this study, accuracy of manufacturing is the main factor influencing the porosity and pore size, eventually changing the mechanical properties and osseointegration. Therefore, to control the processing precision is significant for implantable devices. Powder adhesion is an inevitable problem leading to bad accuracy of pore size and porosity in SLM process. In addition, there is a certain temperature gradient between layer and layer, melting channel and melting channel. This temperature gradient can lead to interior residual stress, eventually results in dimensional difference and even deformation. Therefore, the laser energy input should enough to melt the powderabsolutely, the phenomenon of powder adhesion would be reduced.Also, heat treatment and optimization of technical parameters are used to reduce or eliminate residual stress.

Although the static stress shielding and biocompatibility have shown some improvement with porous structure, the fatigue performance is the most significant factor for implantable devices. Therefore, further study of fatigue testing should be done to evaluate the fatigue performance and feasibility.

5. Conclusion

1. According to the results, with 0.75-mm pore size, the errors of porosity only 5% and the elastic modulus is close to natural bone tissue. Therefore, the pore size of 0.75 is acceptable.
2. The yield strength of this structure is 170.3MPa, which is more than maximum yield stress of bone tissue(165MPa). This structure is reasonable.
3. The stress-strain curve of porous structure by SLM is different from that of common solid structure. Porous structure is with poor ductility.

Since bone replacement biomaterials usually experience cyclic loading, it is important to understand their fatigue behavior and take further study.

Acknowledgments
This work is supported by Science and technology plan program of Guangzhou (No. 2017010160546); Science & Technology Plan Project of Guangzhou (No.201604020147); 2016 International science and technology cooperation project of Nansha;Foshan city cooperation projects (No.2014HT10008).

References

[1] jian-yu Chen zhi-guang zhang, Li Zifu. Selective laser melting technology application in the field of stomatology [J]. International journal of oral medicine, 2014, 01: 97-101.
[2] Attar H, Calin M, Zhang LC, Scudino S, Eckert J. Manufacture by selective laser melting and mechanical behavior of commercially pure titanium. Materials Science & Engineering A 2014;593:170-7.
[3] G. Campoli, M.S. Borleffs, S. Amin Yavari, R.Wauthle, H.Weinans, A.A. Zadpoor, Mechanical propert ies of open-cell metallic biomaterials manufactured using additive manufacturing, Mater. Des. 49 (2013) 957–965.
[4] Ruidi li. The key to the selective laser melting metal powder forming basic question research [D]. Huazhong university of science and technology, 2010 (In chinese).
[5] J. Lee, V. Frias, K. Lee and R. F. Wright, Effect of implant size and shape on implant success rates: A literature review,The Journal of Prosthetic Dentistry, 4 (4) (2004) 377-381.
[6] WuYingYing, GongPing. Research status and progress of the beginning of the implant stability [J]. International journal of stomatology, 2009 practices: 726-728.
[7] Vassilis Karageorgiou, David Kaplan. Porosity of 3D biomaterial scaffolds and osteogenesis, Biomaterials. 2005,26(27):5474-5491
[8] M.T. Andani, N.S.Moghaddam, C. Haberland, D. Dean,M.J. Miller,M. Elahinia, Metals for bone implants. Part 1. Powder metallurgy and implant rendering, Acta Biomater. 10 (10) (2014) 4058–4070.
[9] Hulbert SF, Young FA, Mathews RS, Klawitter JJ, Talbert CD, Stelling FH. Potential of ceramic materials as permanently implantable skeletal prostheses. J BiomedMater Res 1970;4(3):433–56.
[10] E Sallicaleva，AL Jardini，JB Fogagnolo. Microstructure and mechanical behavior of porous Ti－6Al－4V parts obtained by selective laser melting. Journal of the Mechanical Behavior of Biomedical Materials. 2013, 26(30):98-108

IOP Conf. Series: Materials Science and Engineering **212** (2017) 012007 doi:10.1088/1757-899X/212/1/012007

Test-bed's Development of the Packed Bed with Both Revolution&Rotation's High Gravity

Chongpeng ZHANG , Zhiqiang DONG, Bianhua HE

(School of the Mechanical Engineering, Taiyuan University of Science and Technology, Taiyuan 030024, China

654028129@qq.com

Abstract: Rotating packed bed is a chemical reactor which used the effect of rotating centrifugal force to be equivalent in the gravity field. To increase the fluid flow time in bed, to extend the controlled process in bed's reaction and to strengthen the role of influencing factor, a new type of rotating packed bed is proposed in the paper which is running on the revolution platform. Unfortunately, the bad synergy effect of revolution&rotation in packed bed has induced the serious vibration problem. Through optimizing the mass center of packed bed and disposing local weight, the stability of the operation of the bench was more improved. The stability of the operation of the bench was that more improved the bed's vibration amplitude was analyzed by using the ANSYS. The results show that the test-bed of both the Revolution& Rotation's Packed bed is a practical project in the Rotating packed bed's development.

1. Introduction

Rotating packed bed is a chemical reactor which uses the centrifugal effect to enhance fluid flow replacing and overreaching the equivalent gravity field [2, 3].The technology about rotating beds is called the high gravity technology by academician. Its researches mainly focus the following three aspects:①the high gravity of nanometer materials preparation; ②the high gravity of response reinforcement; ③the high gravity of separation technology[4-8]. Its core component is the internal rotor which filled with porous mediums. The chemical fluid flows in the porous medium. The fluid, running through the medium which rotates around rotor with high-speed, is crushed and teared into nanoscale wires, droplets and liquid membranes. Resulting in a huge phase interface which enforces gas-liquid contact and micro-mixing in a great extent [9].A typical High gravity packed bed is show in Fig. 1. The fluid flowing process in bed starts from the liquid distributor through being sprayed into the rotation filler.

Due to the packing is rotating at a high speed, the relative velocity with the liquid phase is very large and when the liquid phase passes through the first few layers of the packing, it was shattered, torn. The liquid phase mass transfer is given priority with dispersion of mass transfer at this stage; the crushed liquid to move forward after being trapped by the filler, and then flow along the packing in the form of brin or liquid film, with the droplets penetrating. This phase is given priority with the forced convection of mass transfer and mass propagation speed, after the liquid phase through the packing, it flowing out from the port d. The gas enter the packing from port c, under pressure to overcome the centrifugal force through the filler retroactively, through contact with the liquid phase interface to carry out the chemical reactions, then flowed out from port b. [11]

The three kinds of mass transfer in the high gravity packed bed are dispersion of mass transfer, the forced convection of mass transfer and mass propagation speed. The interaction in this Three-scale mass transfer determines the effect of chemical reaction. The effect of mass transfer is the largest among the three, and it is also the basis of convective mass transfer and mass propagation speed. Exist the following problems, problem 1: Dispersion of the mass transfer area is very thin, the reason is that the liquid phase is completely captured in several layers that are very thin in the front region of the packing, the relative velocity of fluid and filler is zero, and effect of dispersion disappear. In particular, more than two components of the fluid bed in the packed bed are required to have a higher degree of dispersion and mass transfer, resulting in a more severe problem. Problem 2: the limited thickness of the filler layer directly determines the course of liquid phase through the filler is too short. The process is too short and shortens the time of controllable chemical reaction, and affecting the final response. Such as method of gas-liquid based on high gravity packed bed to preparation of nanoparticles, etc.

Based on this, a new type of the Revolution& Rotation's Packed bed is proposed, which is based on the introduction of rotating system on the basis of the rotating packed bed and realize the continuous production

2. The composition of test-bed

2.1 The rotation orbit High gravity the general structure of the rotating packed bed

The packed bed comprises a feed system, a revolution system, a rotation system and a discharge system. This paper increase one level of degree of rotational freedom in the filler area, change it into coexistence of revolution and rotation, so that we can rapidly change the direction of High gravity over time. One of the effect is effectively expanding the radial dispersion of the filler to the liquid phase, and enhance the effect on dispersion of mass transfer; second increase the total process of liquid phase in filler, and enhance the effect of integrated mass transfer.

In the Revolution&Rotation's Packed bed, the relative velocity of the fluid and the filler is increased by increasing the rotation of the reaction kettle, and expanded the course of liquid phase through the filler. In this study, the experimental platform of nanometer calcium carbonate was designed and prepared to predict the effect of the mass transfer. In order to promote the experimental work in a phased manner, the continuous pipeline composed of feeding and receiving system is simplified. The feeding box and the receiving box are fixed on the revolution platform, and the experimental platform is constructed.

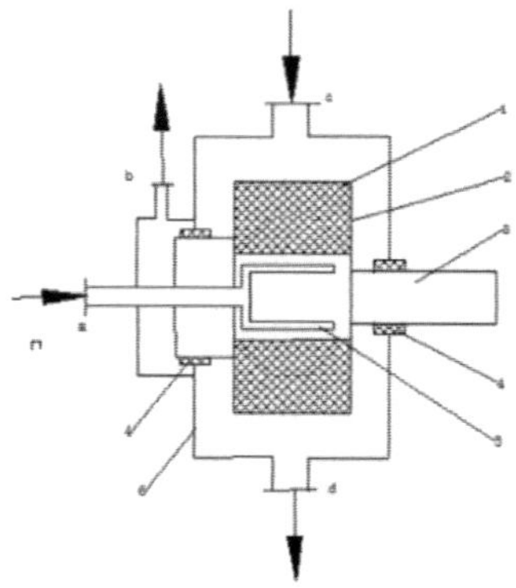

1 padding 2 otor 3 spindle 4 sealing gasket 5 The cloth liquid 6 shell a liquid inlet b gas outlet c gas inlet

Figure 1. The traditional high gravity packed bed structure diagram

Figure 2. The physical Fig. of revolution&rotation high gravity packed bed

This experiment at the core of the reaction kettle using circular structure

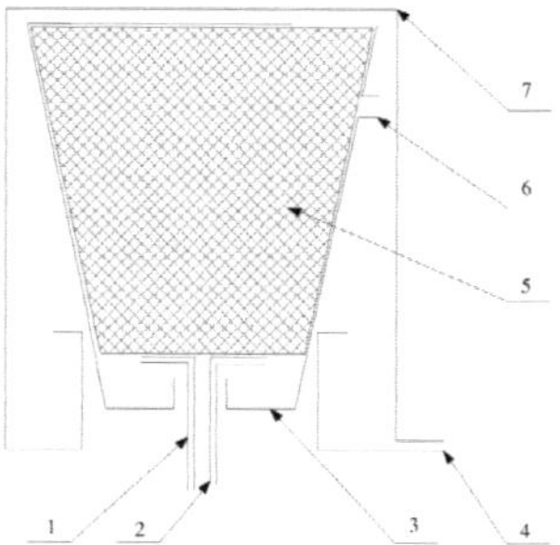

1 liquid inlet tube 2 intake-tube 3 Outside shell of the reaction kettle 4 discharge hole 5 Porous media filler 6 the discharge 7 hole of the reaction kettle shell 8 Outside shell of the reaction kettle

Figure 3. The Schematic diagram of the reaction kettle

3. IMPROVEMENTS ON TEST-BED

In the first test, the test-bed has a large noise and swing seriously affected the experiment and the life of test-bed, and makes the speed ratio of rotation and revolution cannot be controlled, largely affected the experiment. The analysis of the reasons resulting from such a result may have the following aspects. First, natural frequency of the test-bed and excitation frequency may resonate; Second, the reason for the swing may be that the amplitude of the mass center of the test-bed is too large. In the actual construction of the test-bed, many sizes of raw materials, relative position are difficult to balance. Therefore, this paper improve the test-bed from both the optimizing the mass center of packed bed and disposing local weight. This paper used simulation software to test the effect of the improvement.

3.1. Modal analysis was carried out on the test-bed

The layout of the test-bed is shown in the following Fig., Fig. 4 is the original layout, and Fig. 5 is the optimized layout. The black part is a clump weight

Because the mass distribution of the objects on the test-bed is not uniform, the mass center of the test-bed is not in the geometric center. As a result, the table turns unstable. To make the mass center of the test-bed in the geometric center, the sum of the centrifugal force in all directions should meet the following formula.

$$\sum_{i=1}^{n} m_i x_i \omega^2 = 0 \tag{1}$$

$$\sum_{i=1}^{n} m_i y_i \omega^2 = 0 \tag{2}$$

$$\sum_{i=1}^{n} m_i z_i \omega^2 = 0 \tag{3}$$

Where m_i is the weight of each component, x_i, y_i, z_i is the mass center coordinates of each component, ω the angular velocity .

According to the actual situation to optimize the layout, take the lowest center of the axis as the origin, Solidworks can be used to calculate the location of the center of mass

$(x, y, z) = (614.95, 588.16, 749.39)$ All the objects on the test-bed equivalent to an object M, it can know the quality of all objects on the test-bed M = 190895.11 g

$$Mx\omega^2 + m_a x_a \omega^2 = 0 \tag{4}$$

$$My\omega^2 + m_a y_a \omega^2 = 0 \tag{5}$$

$$Mz\omega^2 + m_a z_a \omega^2 = 0 \tag{6}$$

Where m_a is the quality of the weight, x_a, y_a, z_a the coordinates of weight.

Combined with the actual situation of the layout shown in Fig. 5. In Fig. 4 and Fig. 5, the red coordinates is the position of the mass center, and it can be seen that the optimized layout is closer to the geometric center of the test-bed.

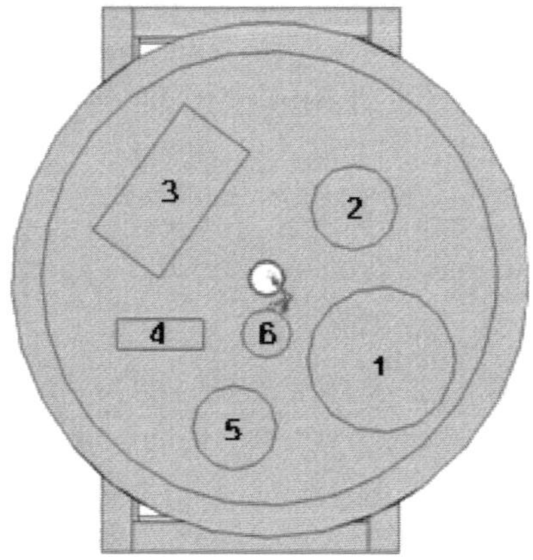

1 the reaction kettle 2 the electromotor 3 feeding equipment and receiving device 4 console 5 transducer 6 CO_2 battery

Figure 4. The original layout

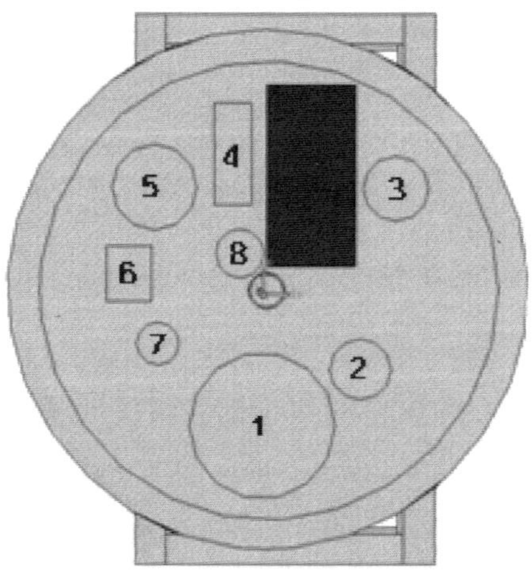

1 the reaction kettle 2 the incoming container 3 the transducer 4 the receiving device 5 the electromotor 6 the feeding equipment 7 the receiving bottle

Figure 5. The improved layout

3.2 The experimental modal analysis

As an auxiliary method in structural design, modal analysis can be used to determine the vibration frequency and mode of the structure, and to predict the influence of the external excitation frequency on the structure, so as to avoid the resonant frequency during the design process. Modal analysis on the overall structure of the test-bed can be carried out to analyze the influence of external excitation on the structure of the test-bed and further optimizes the structure of the test-bed. In this paper, the modal analysis of the test-bed is carried out, and the first 10 order vibration frequencies and the main vibration modes of the test-bed are obtained, and the results are used to provide the theoretical support for the

IOP Conf. Series: Materials Science and Engineering **212** (2017) 012007 doi:10.1088/1757-899X/212/1/012007

optimization design of the test-bed.

Using SOLIDWORKS to builds the three-dimensional geometry model of the test-bed, the original layout of the model into ANSYS for meshing. The meshes of the test-bed are triangular meshes, the number of nodes is 64041 and the number of units is 192060. The modal analysis in ANSYS is a linear analysis that requires specifying the modulus of elasticity EX and the density DENS. If there are nonlinear elements, default that it is linear. The model adopts high-order element entity, the material adopts Q235A, the thickness is 5mm, the density is 7.86g / cm^3, the elastic modulus is 210Gpa, the Poisson's ratio is 0.3, and block lanczos modal extraction method is adopted.

3.3 modal analysis test-bed base

ANSYS was used to obtain the first 10 order modal frequencies of the test bed. As shown in Table 1 and Fig. 6.

Table.1 Modal frequency

Order/n	Modal frequency f/（Hz）
1	66.020
2	75.278
3	85.805
4	86.690
5	90.232
6	111.55
7	138.88
8	139.34
9	153.07
10	179.40

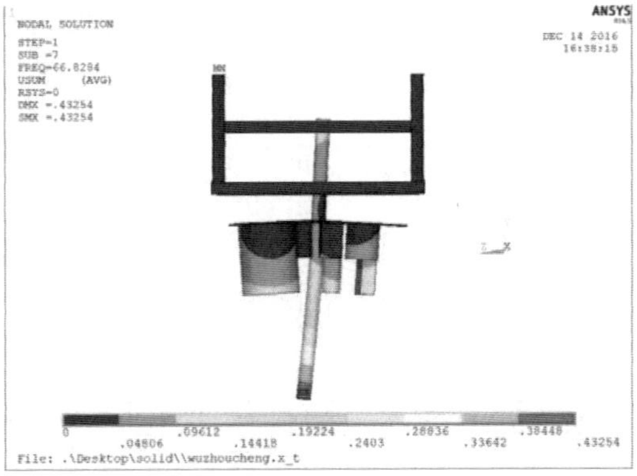

Figure 6. The first step model of the former improved test-bed

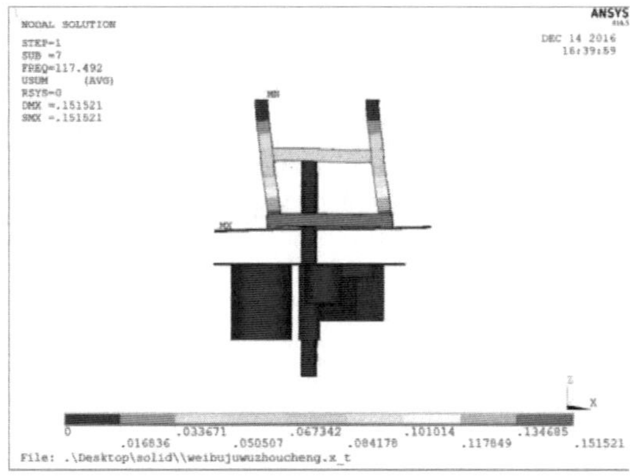

Figure 7. The first step model of the improved test-bed

When the natural frequency of the test-bed close to excitation frequency, it's easy to produce resonance. Therefore, it is better to ensure the excitation frequency is lower than the lowest natural frequency of the test-bed during the experiment, and the lowest natural frequency of the test-bed is more than 1.25 times to the maximum frequency. It is known that the maximum speed of the motor is 1400r / min and the maximum frequency is 50Hz. In this speed range, the lowest order natural frequency of the test-bed is 1.25 times to the maximum excitation frequency of the motor, so as to avoid its influence on the experimental study [12].

The speed of the test-bed is controlled by the transducer controlling the motor, and the excitation frequency of base of the test-bed is also changed. The excitation frequency at different speeds can be obtained from the following formula:

$$f=\frac{n}{60}\times\frac{P}{2} \qquad (7)$$

where n is the motor speed, 2p is logarithmic.

The natural frequency of the test-bed is calculated to be at least 62.5 Hz, while the first-order modal frequency of the test-bed is 66.020 Hz, which is close to that and should be improved.

The optimized first 10 modes are shown in Table 2 and Fig. 7

Table.2 Modal frequency

Order/n	Modal frequency f/（Hz）
1	117.49
2	122.07
3	143.34
4	181.55
5	197.26
6	212.00
7	233.87
8	296.51
9	299.48
10	333.18

It can be seen that the improved mode has a larger enhancement compared with the original layout mode, thus keep away from the excitation frequency.

4. Conclusion

In this paper, the construction of a new type of high-gravity rotating bed is introduced. Considering the vibration of the test-bed is too large, it is improved from the optimizing the mass center of packed bed and disposing local weight, and the modal analysis was carried out before and after the optimization. It shows that the natural frequency of the improved test-bed avoids the influence of the excitation frequency and does not produce resonance phenomenon. Taking into account the swing of the test-bed, Adams analysis was performed before and after optimization of the test-bed, and the analysis results show the mass center amplitude of the improved test-bed. The later experiments also proved this point.

Reference

[1] Chen Jianfeng, Guo Kai, Guo Fen,et al. The Higee Technology and Application: A New Generation of Reaction and Separation Technology [M]. Beijing: Chemical Industry Press, 2003

[2] Institute of Process Engineering, Chinese Academy of Sciences. Process engineeringmaterial • energy • wisdom [M]. Beijing: Science Press, 2010.410 -449.

[3] Chu Guangwen,Luo Yong,Zou Haikui, et al. Reaction process intensification: application of acidic tail-gas treatment by high gravity technology [J]. Chemical Reaction Engineering , 2013, 29 (3): 193-198

[4] Gao Xin.Studies of distillation in a counter-current rotating packed bed [D]. Beijing of Chemical Technology, 2010

[5] Sang L, Luo Y, Chu G W, et al. Liquid flow pattern transition, droplet diameter and size distribution in the cavity zone of a rotating packed bed: A visual study[J].

[6] Chemical Engineering Science, 2016, 158:429-438.

[7] Zhang J,Gao D,Li Y, et al. Research progress of multiphase transport in high gravity environment in rotating packed bed[J]. Ciesc Journal, 2013. 64 (1): 243-251.

[8] Zhao H, Shao L, Chen J F. High-gravity process intensification technology and application [J]. Chem. Eng. J., 2010, 156 (3):588-593.

[9] Chen J F, Wang Y H, Guo F, Wang X M, Zheng C. Synthesis of Nanoparticles with novel technology: high-gravity reactive precipitation [J]. Ind. Eng. Chem. Res., 2000, 39 (4): 948-954.

[10] Rao D P, Bhowal A, Goswami P. Process intensification in rotating packed beds (HIGEE): an appraisal [J]. Ind. Eng. Chem. Res., 2004,43 (4): 1150-1162.

[11] Chen J F, Shao L. Mass production of nanoparticles by high gravity reactive precipitation technology with low cost [J]. China Part.,2003, 1 (2): 64-69.

[12] Zhang liangliang. study on intensification of decarbonization and denitration by absorption using rotating packed bed[D];Beijing University of Chemical Technology

IOP Conf. Series: Materials Science and Engineering **212** (2017) 012008 doi:10.1088/1757-899X/212/1/012008

The Influencing Factor Analysis on the Performance Evaluation of Assembly Line Balancing Problem Level 1 (SALBP-1) Based on ANOVA Method

Jie Chen [1], **Jiangnan Hu** [2,*]

[1, 2]School of Economics and Management, Nanjing University of Science and Technology, 210094, China

jiangnan317@outlook.com

Abstract. Industry 4.0 and lean production has become the focus of manufacturing. A current issue is to analyse the performance of the assembly line balancing. This study focus on distinguishing the factors influencing the assembly line balancing. The one-way ANOVA method is applied to explore the significant degree of distinguished factors. And regression model is built to find key points. The maximal task time (t_{max}), the quantity of tasks (n), and degree of convergence of precedence graph ($conv$) are critical for the performance of assembly line balancing. The conclusion will do a favor to the lean production in the manufacturing.

1. Introduction

In 1914, the assembly line born in Ford has greatly improved labor productivity. Mass production has been allowed under the assembly line cell. The manufacture focus on efficiency and reducing waste increasingly, more and more manufacturing enterprises push lean production.

However, there are stillsome waste in assembly line, and evaluating the performance of assembly line balancing is critical to reduce the waste, realize lean production and maximize the value in manufacture period. Identifying the factors influencing the assembly line balancing performance is benefit for the design of assembly cell. ANOVA and regression analysis have been considered by introducing statistical methods.

We consider the simple assembly line which continuously manufactures a homogeneous product in mass production. The simple assembly line balancing problem level 1 (SALBP-1, see [1] for some definitions and terminology) is to find an optimal balance under a given cycle time c, and seeking a feasible assignment of the given operations to stations contribute to the minimal of m (the number of the stations). Generally, the SALBP-1 is formulated as follows:

Given $N(1,2,...,j,...,n)$ operations and $K(1,2,...,k)$ stations, and the process time t_j for each operations is defined. And the cycle time c is defined as the time span between two productions produced continuously. The formulation (1) as follows could explain the definition adequately.

$$c = \textit{The effective time in the planning period} / \textit{The output in the planning period} \quad (1)$$

Moreover,finish-start precedence relations $i \rightarrow j$ are defined between the operations according to anacyclic directed graph G [2]. Table 1 shows the details of notations.

For the precedence relationship between the operations, the precedence graph (an example, figure.1) shows it well, and the circles express the operations in an assembly line.

Table1. Notations of SALBP-1.

Notation	Connotation	**Notation**	Connotation
c	Cycle time	t_j	Process time of operation j
j	Index of operation	t_{min}	The minimal process time of operation
k	Index of station	t_{max}	The maximal process time of operation
m	Number of stations	t_{sum}	The total process time of operation
n	Number of operations	tv	Time variation proportion

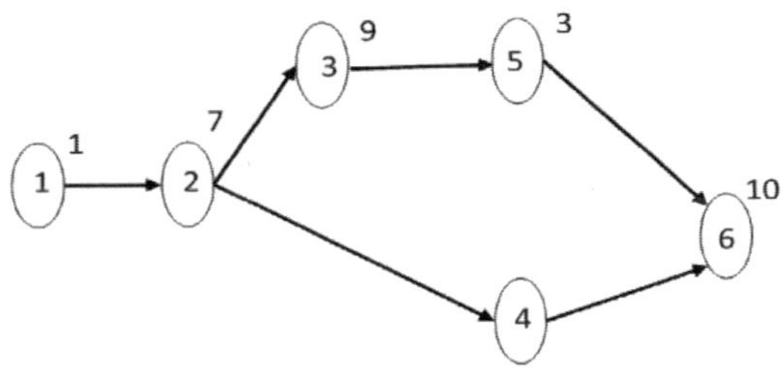

Figure 1. An example of precedence graph.

2. Determining the Index of Performance Evaluation

The assembly line balancing is not only a critical problem in academia,but also an increasingly growing focus of actual industry. There are three indexes for the SALBP-1.

2.1. Index Establishment

The first index which indicates the performance of the SALBP-1 is balance delay (BD), in other words, the time loss coefficient. And BD represents the percentage between the total free time which resulted by producing one product and the total assembly time in an assembly line. Balancing delay, precisely for the allocation of tasks on the assembly line to stations is not evenly, resulting in free time. The formula (2) shows the connotation in detail and table 2 shows the performance evaluation criteria.

$$BD = 1 - t_{sum} / (m^* \times c) \quad (2)$$

Table 2. The performance evaluation criteria.

	BD≦10%	10%<BD<20%	BD≧20%
Performance of SABLP-1	excellent	good	bad

The second index is balance efficiency (BE), which showed in formula (3).

$$BE = t_{sum} / (m^* \times c) \quad (3)$$

In addition to BD and BE ,another index is called load ratio of station, which mainly set the station of the operation's longest time as a standard. Considering BD , BE and load ratio, we set the observed variables as $SALBE$ (m^* is the optimal number of stations in theory), being represented as follows.

$$SALBE = t_{sum} / (m^* \times c) \quad (4)$$

$$m^* = \left[\sum t_j / c \right] \quad (5)$$

2.2. Identifying the Factors

For better balance performance, the main factors influencing SALBP-1 will be pursued in this paper.

2.2.1. Cycle Time

Cycle time (c) is the detention period of the products, namely, the time between one product and the next one. And thus the cycle time determines the production capacity of assembly line in a certain period. Regardless of the relation between BE and c, the cycle time may also has effects on the minimum number of stations. Thus we cannot get the specific influence between BE and c.

2.2.2. Process Time of Operation

The process time of operation may have influence on balance delay and balance efficiency. Also, the minimal process time of operation (t_{min}), maximal process time of operation (t_{max}) and total process time of operation (t_{sum}) may have influence on the assembly line balancing. And in this paper, we use the rate of time change (tv) to analyze the influence effect.

$$tv = t_{max} / t_{min} \tag{6}$$

2.2.3. Number of Operations

The larger the number of operations is, the more complicated the assembly line will be, and the solutions to balancing will be more complicated. Then, the number of operations (n) is another factor.

2.2.4. Precedence Related

The precedence graph shows the precedence relationship between operations.

Order strength (os), and its practical significance is, in an algorithm case of assembly line balancing problem level 1, reflecting the complexity of the algorithm when solving the SALBP-1.The remaining two factors are about the two aspects of precedence graph' convergence. The notation div represents the degree of divergence of precedence graph. The notation $conv$ is the degree of convergence of precedence graph.In summary, there are nine factors ($c, t_{min}, t_{max}, t_{sum}, tv, n, os, div, conv$) which may influence the performance in the SALBP-1.

3. The ANOVA Model

To analyse whether there are significant influence to the performance of SABLP-1 (we setthe observed variables as $SALBE$),the single control variables including $c, t_{min}, t_{max}, t_{sum}, tv, n, os, div, conv$,the single factor analysis of variance model has been built as follows:

$$x_{ij} = \mu + \alpha_i + \varepsilon_{ij}, (i=1,2, \ldots ,k; j = 1,2, \ldots ,r) \tag{7}$$

It's important to be noticed that α_i reflects that the additional effect of control variable's level A_i influencethe observed variable. And k represents the number of control variable's level; r shows the number of sample in each control variable's level; μ is the grand sum of observed variable in A_i, and $\mu = \mu_i - \alpha_i$; then, ε_{ij} typifies the error of sample (which subject to the standard normal distribution) ; for the x_{ij} , the value of sample j of A_i.

To findwhether the balance performance, which is influenced by control variables, is significant or not, the proof-test being detected whether all effects of control variable is 0 at the same time.Then set up the null hypothesis that different levels of control variables has no significant influence to the observed variable , $\alpha_i = 0, i = 1,2,\ldots,k$ .To verify the null hypothesis,we useF-statistics.

$$\text{F} = \frac{(n-k) SS_{interclass}}{(k-1) SS_{intraclass}} = \frac{MS_{interclass}}{MS_{intraclass}} \sim F(k-1, n-k) \tag{8}$$

For the details of terms presented in formula (8), $SS_{interclass}$ isthe sum of squares of deviations in interclass, $MS_{interclass}$ represents the average for the sum of squares of deviations in interclass, $SS_{intraclass}$ and $MS_{intraclass}$ represent the same value in intraclass respectively.

3.1. The Scatter Diagram of Takt Time

It's no doubt that the performance of assembly line balancing is influenced by the takt time c. Two algorithm cases (Arcus1 and Scholl) are chosen to analysis the influence to $SALBE$.

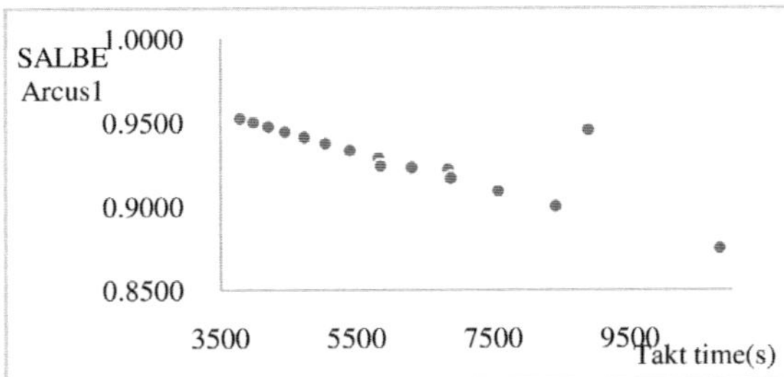

Figure2. The scatter diagram of Arcus1.

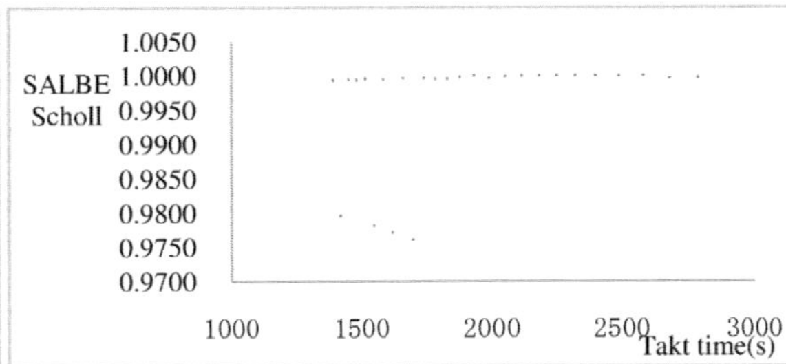

Figure3. The scatter diagram of Scholl.

Figure 2 shows negative correlation between c and $SALBE$ of case Arcus1.Figure 3 shows no significant relationship in Scholl.Takt time influences $SALBE$ depending on different algorithm cases.

3.2. The ANOVA Analysis of Other Factors

The sample data for this study are selected by setting nearly 10 stations from algorithm cases such as Hahn, Heskiaoff, Jackson, Jaeschke, Kilbridge, Lutz1(reference to [3]-[7]).

Table 3. The homogeneity test of variance.

Factor	Levene statistic	df1	df2	Significant
t_{sum}	1.444	9	49	0.196
tv	2.049	7	51	0.067
n	1.469	8	50	0.193
os	1.444	9	49	0.196
$conv$	1.547	7	51	0.173
t_{max}	2.049	7	51	0.067
t_{min}	6.127	4	54	0.000
div	1.656	8	50	0.133

The single factor analysis of variance ANOVA is conducted by SPSS software. Firstly, conducting the homogeneity test of variance by the method of homogeneity of variance, and the results are listed in Table 3.The result shows $t_{max}, t_{sum}, tv, n, os, div, conv$ meet the critical value 0.05.Thus these 7 factors are homoscedasticity except t_{min}. The single factor analysis of variance's results are showed in Table 4.

Table 4. One-way ANOVA results.

Factor	F	Sig.	$MS_{interclass}$	$MS_{intraclass}$
t_{sum}	5.841	0.000	0.260	0.004
tv	6.989	0.000	0.032	0.005
n	4.691	0.000	0.024	0.005
os	5.841	0.000	0.026	0.004
$conv$	7.056	0.000	0.032	0.005
t_{max}	6.989	0.000	0.032	0.005
div	6.703	0.000	0.030	0.004

Table 4 reflects that the average for the sum of squares of deviations in interclass and interclass discrepancy mainly from interclass.F-statistics>1, reflects the influence to $SALBE$ is significant.

4. The Stepwise Regression Model

Stepwise regression analysis model is built to determine the key factors. Stepwise regression analysis is conducted in SPSS. Set the standard of step by step method as follows: the probability of entering is 0.05, the probability of eliminating is 0.10. Table 5 shows the results of stepwise regression analysis.

Table 5. Stepwise regression analysis.

Model		Unstandardized Coefficient	Standardized Coefficient	F	t	Sig.	R^2
1	Constants	1.015		0.001	28.56	0.000	0.16
	os	-0.002	-0.422		-3.51	0.001	
2	Constants	0.973		0.000	24.72	0.000	0.22
	os	-0.002	-0.412		-3.54	0.001	
	n	0.001	0.253		2.71	0.034	
3	Constants	0.898		0.000	17.44	0.000	0.26
	os	-0.001	-0.202		-1.36	0.180	
	n	0.003	0.476		3.13	0.003	
	t_{max}	0.000	-0.388		-2.18	0.034	
4	Constants	0.834		0.000	40.34	0.000	0.25
	n	0.003	0.571		4.19	0.000	
	t_{max}	0.000	-0.545		-4.00	0.000	
5	Constants	0.330		0.000	2.21	0.032	0.37
	n	0.006	0.942		5.67	0.000	
	t_{max}	0.000	-0.715		-5.31	0.000	
	conv	0.622	0.470			0.001	

The result shows the coefficient of determination of the regression model is 0.37. In a certain extent, the key factors influencing the assembly line balance's performance are: the number of operations, the longest time of operations and the degree of convergence of precedence graph.

5. Conclusion

This research mainly presents two contributions to manufacturing. Firstly seven factors which have significant influence to the performance of the assembly line balancing are found based on the sample data with the ANOVA model. The additional contribution is that the key factors influencing the first type assembly line balancing efficiency are verified through the stepwise regression analysis model. As for the conclusion by applying theanalysis of variance and regression analysis, there are two suggestions for the assembly line manufacture on two sides. More focus need to be paid for the degree of convergence of precedence graph and the number of operations in finding the optimal solution of SALBP-1. More training for the bottleneck operation should be organized to reduce the longest task time.

References

[1] Erdal Erel, Subhash C. Sarin. A survey of the assembly line balancing procedures [J]. Production Planning Control, 1998, 9(5):414-434.
[2] Gafarov E R, Dolgui A. Hard Special Case and Other Complexity Results for SALBP-1[J]. 2012.
[3] Nearchou A C. A differential evolution algorithm for simple assembly line balancing[C]//In16th International federation of automatic control (IFAC) World Congress. Prague. 2005.
[4] Scholl A, Klein RSALOME: A bidirectional branch-and-bound procedure for assembly line balancing[J]. INFORMS Journal on Computing, 1997, 9(4): 319-334.
[5] Fleszar KHindi K S. An enumerative heuristic and reduction methods for the assembly line balancing [J]. European Journal of Operational Research, 2003, 145(3): 606-620.
[6] Pape T. Heuristics and lower bounds for the simple assembly line g problem type 1: Overview,

computational tests and improvements [J]. European Journal of Operational Research, 2015, 240(1):32-42.

[7] García-Villoria A, Corominas A, Pastor R. Heuristics and simulated annealing procedures for the accessibility windows assembly line problem level 1 (AWALBP-L1)[J]. Computers & Operations Research, 2015, 62:1-11.

IOP Conf. Series: Materials Science and Engineering **212** (2017) 012009 doi:10.1088/1757-899X/212/1/012009

Optimal Trajectories Generation in Robotic Fiber Placement Systems

Jiuchun Gao[1,*], Anatol Pashkevich[1], Stéphane Caro[2]

[1]IMT Atlantique Engineering School, Laboratory LS2N, France
[2]National Center for Scientific Research CNRS, Laboratory LS2N, France

[*]Email: jiuchun.gao@ls2n.fr

Abstract. The paper proposes a methodologyfor optimal trajectories generationinrobotic fiber placement systems. A strategy to tune the parameters of the optimization algorithm at hand is also introduced. The presented techniquetransforms the original continuous problem into a discrete one where the time-optimal motions are generated by using dynamic programming. The developed strategy for the optimization algorithm tuning allows essentially reducing the computing time and obtaining trajectories satisfying industrial constraints. Feasibilities and advantages of the proposed methodology are confirmed by an application example.

1. Introduction

Robotic fiber placement technology has been increasingly implemented recently in aerospace and automotive industries for fabricating complex composite parts[1, 2]. It is a specific technique that uses robotic workcell to place the heated fiber tows on the workpiece surface[3]. Corresponding robotic systems usually include a 6-axis industrial robot and a one-axis positioner (see Figure 1), which are kinematically redundant and provides the user with some freedom in terms of optimization of robot and positioner motions.

To deal with the robotic system redundancy, a common technique based on the pseudo-inverse of kinematic Jacobian is usually applied. However, as follows from relevant studies, this standard approach does not satisfy the real-life industrial requirements of the fiber placement[4, 5]. In literature, there is also an alternative technique (that deals with multi-goal tasks) that is based on conversion of the original continuous problem to the combinatorial one[6, 7], but itonly generates trajectories for *point-to-point* motions, e.g.for spot welding applications.Aslightly different method was introduced in [8-10], andit was successfully applied tolaser-cutting and arc-weldingprocesses where the tool speed was assumed to be constant(which is not valid in the considered problem).Another approach has been proposed in [11], where the authors concentrated on the tool path smoothing in Cartesian spacein order to decrease the manufacturing time in fiber placement applications. For the considered process, where the tool speed variations are allowed (in certain degree), a discrete optimization based methodology was proposed in our previous work[12]. It allows the user to convert the original continuous problem to the combinatorial one taking into account particularities of the fiber placement technology and to generate time-optimal trajectories for both the robot and the positioner.Nevertheless, there are still a number of open questions related to selection of the optimization algorithm parameters (i.e., its "tuning") that are addressed in this paper and targeted to the improvement of the algorithm efficiency and the reduction of the computing time.

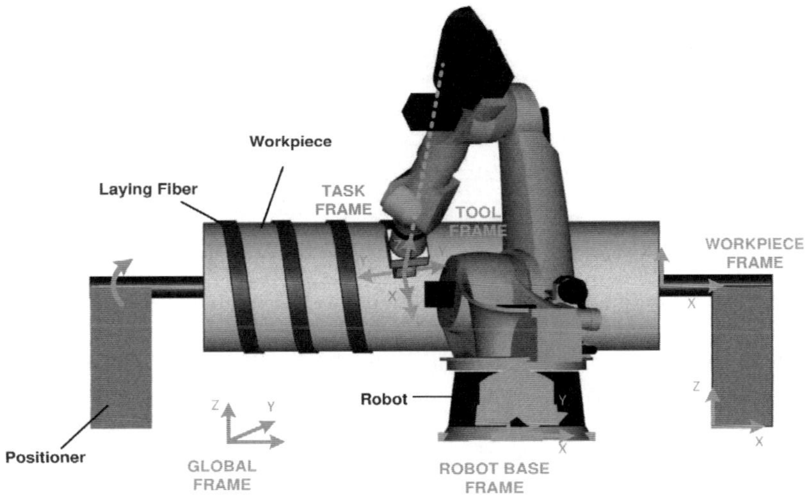

Figure 1. Typical robotic fiber placement system (6-axis robot and one-axis positioner).

2. Robotic system model

In practice, the procedure of off-line programming for robotic fiber placement is implemented in the following way. The fiber placement path is firstly generated and discretized in CAM system. Further, the obtained set of task points is transformed into the task graph that describes all probableconfigurations of the robot and the positioner joints. The motion generator module finds the optimal trajectories that are presented as the "best" path on the graph. Finally, the obtained motions are converted into the robotic system program by the post processor. The core for the programming of this task is a set of optimization routines addressed in this paper.

To describe the fiber placement task, let us present it as a set ofdiscrete task frames $F_{task}^{(i)}, i=1,2,...n$, in such a way that the X-axis is directed along the path direction and Z-axis is normal to the workpiece surface pointing outside of it (see Figure 1). Using these notations, the task locations can be described by 4×4 homogenous transformation matrices and the considered task is formalized as follows:

$$ {}^{w}\mathbf{T}_{task}^{(1)} \rightarrow {}^{w}\mathbf{T}_{task}^{(2)} \rightarrow ... {}^{w}\mathbf{T}_{task}^{(i)} \rightarrow ... {}^{w}\mathbf{T}_{task}^{(n)} \tag{1} $$

where all vectors of positions and orientations are expressed with respect to the workpiece frame (see superscript "w"). To execute the given fiber placement task, the robot tool must visit the frames defined by (1) as fast as possible.

The considered robotic system, shown in Figure 1, is composed of an industrial robot and an actuated positioner. Their spatial configurations can be described by the joint coordinates $\mathbf{q}_R$ and q_P respectively.The task frames can be presented in two ways using the robot and positioner kinematics that are expressed as $g_R(\mathbf{q}_R)$ and $g_P(q_P)$, respectively.To obtain the kinematic model of the whole system that is expressed as a closed loop containing the robot, the workpiece and the positioner, a global frame F_0 is selected. Then,the tool frame F_{tool}, and task frame $F_{task}^{(i)}$ can be aligned in such a way that: (i) the origins of the two frames coincide; (ii) Z-axes are opposite; (iii) X-axes have the same direction. Due to the foregoing closed-loop, two paths can be followed to express the transformation matrices from the global frame to the task frames, namely,

$$ {}^{0}\mathbf{T}_{Rbase} \cdot g_R(\mathbf{q}_R^{(i)}) \cdot {}^{Tool}\mathbf{T}_{task} = {}^{0}\mathbf{T}_{Pbase} \cdot g_P(q_P^{(i)}) \cdot {}^{w}\mathbf{T}_{task}^{(i)};\ i=1,2,...n \tag{2} $$

Equation (2) does not lead to a unique solution for $\mathbf{q}_R$ and q_P as the robotic system, i.e., robot and positioner, is kinematically redundant. Therefore, the optimum robot and positioner configurations can be searched based on specific criteria.

3. Algorithm for trajectories generation

To take advantage of the kinematic redundancy, it is reasonable to partition the desired motion between the robot and the positioner ensuring that the technology tool executes the given task with smooth motion as fast as possible.

To present the problem in a formal way, let us define the functions $\mathbf{q}_R(t)$ and $q_P(t)$ describing the robot and positioner motion as a function of time $t \in [0,T]$. Additionally, a sequence of time instants $\{t_1, t_2, \ldots t_n\}$ corresponds to the cases where the tool visits the locations defined by (1), and $t_1 = 0, t_n = T$. As a result, the problem at hand is formulated as an optimization problem aiming at minimizing the robot processing time

$$T \rightarrow \min_{\mathbf{q}_R(t), q_P(t)} \tag{3}$$

This problem is subjected to the equality constraints ${}^0\mathbf{T}_{Rbase} \cdot g_R(\mathbf{q}_R(t_i)) \cdot {}^{Tool}\mathbf{T}_{task} = {}^0\mathbf{T}_{Pbase} \cdot g_P(q_P(t_i)) \cdot {}^w\mathbf{T}^{(i)}_{task}$ defined in (2) and some inequality constraints associated to the capacities of the robot/positioner actuators that are defined byupper bounds of the joint velocities and accelerations. Besides, the collision constraints verifying the intersections between the system components are also taken into account.

For this considered problem aiming at finding desired continuous function of $\mathbf{q}_R(t)$ and $q_P(t)$, there is no standard approach that can be applied to straightforwardly. The main difficulty here is that the equality constraints are written for the unknown time instants $\{t_1, t_2, \ldots t_n\}$. Besides, this problem is nonlinear and includes a redundant variable. For these reasons, this paper presents a combinatorial optimization based methodology to generate the desired trajectories.

For the considered robotic system, there is one redundant variable with respect to the given task. It is convenient here to treat q_P as the redundant one since it allows us to use the kinematic models of the robot and the positioner independently and to consider the previous equality constraints.

To present the problem in a discrete way, the allowable domain of $q_P \in [q_P^{\min}, q_P^{\max}]$ is sampled with the step Δq_P as $q_P^{(k)} = q_P^{\min} + \Delta q_P \cdot k; k = 0,1,\ldots m$, where $m = (q_P^{\max} - q_P^{\min})/\Delta q_P$ .Then, applying sequentially the positioner direct kinematics and the robot inverse kinematics, a set of possible configuration states for the robotic system can be obtained as $\mathbf{q}_R^{(k)}(t_i) = g_R^{-1}({}^{Rbase}\mathbf{T}_0 \cdot {}^0\mathbf{T}_{Pbase} \cdot g_P(q_P^{(k)}(t_i)) \cdot {}^w\mathbf{T}^{(i)}_{task} \cdot {}^{task}\mathbf{T}_{tool}, \mu); i = 1,2,\ldots n$, where $\boldsymbol{\mu}$ is a configuration index vector corresponding to the robot posture. Therefore, for i^{th}task location, a set of candidate configuration states can be obtained, i.e., ${}^w\mathbf{T}^{(i)}_{task} \rightarrow \mathbf{L}^{(k;i)}_{task}; \forall k$, where $\mathbf{L}^{(k;i)}_{task} = (\mathbf{q}_R^{(k)}(t_i), q_P^{(k)}(t_i))$.

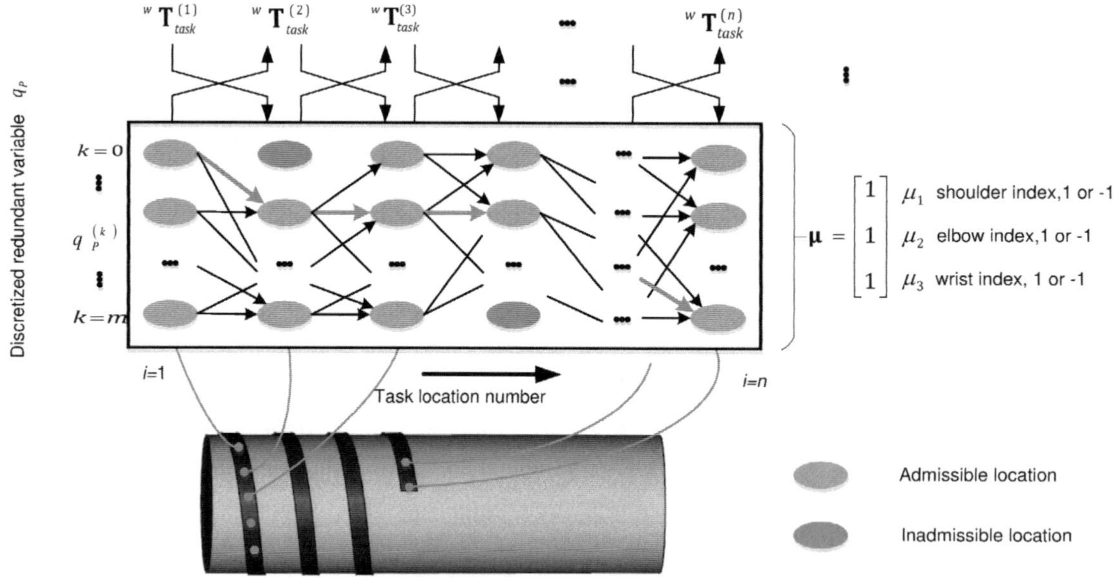

Figure 2. Graph-based representation of the discrete search space

After presenting ${}^{w}\mathbf{T}_{task}^{(i)}$ in joint space as above, the original task can be converted into the directed graph shown in Figure 2. It should be noted that some of the configuration cells should be excluded because of violation of the collision constraints or the actuator joint limits. These cases are denoted as "inadmissible" in Figure 2, and are not connected to any neighbor. Here, the allowable connection between the graph nodes are limited to the subsequent configuration states $\mathbf{L}_{task}^{(k',i)} \rightarrow \mathbf{L}_{task}^{(k',i)}$, and the edge weights correspond to the minimum robot processing time restricted by the maximum velocities and accelerations of the robot and the positioner.

Using the discrete search space above, the considered problem is transformed to the classic shortest path searching and the desired solution can be represented as the sequence $\{\mathbf{L}_{task}^{(k_1,1)}\} \rightarrow \{\mathbf{L}_{task}^{(k_2,2)}\} \rightarrow ...\{\mathbf{L}_{task}^{(k_n,n)}\}$. The objective function (robot processing time)can be presented as the sum of the edge weights

$$T = \sum_{i=1}^{n-1} dist(\mathbf{L}_{task}^{(k_i,i)}, \mathbf{L}_{task}^{(k_{i+1},i+1)}) \tag{4}$$

that depends on indices $k_1, k_2, ...k_n$, where $dist(\mathbf{L}_{task}^{(k_i,i)}, \mathbf{L}_{task}^{(k_{i+1},i+1)}) = \max(|q_{j,i}^{(k_i)} - q_{j,i+1}^{(k_{i+1})}| / \dot{q}_j^{\max}); j = 0,1,...6$. It should be mentioned that the above expressiontakes into account the velocity constraints automatically and the acceleration constraints should be considered by means of the following formula:

$$2\left|\Delta t_i (q_{j,i+1}^{(k_{i+1})} - q_{j,i}^{(k_i)}) - \Delta t_{i+1} (q_{j,i}^{(k_i)} - q_{j,i-1}^{(k_{i-1})})\right| / \Delta t_{i+1} \Delta t_i (\Delta t_{i+1} + \Delta t_i) \leq \ddot{q}_j^{\max} \tag{5}$$

where $\Delta t_{i+1} = dist(\mathbf{L}_{task}^{(k_i,i)}, \mathbf{L}_{task}^{(k_{i+1},i+1)})$ and $\Delta t_i = dist(\mathbf{L}_{task}^{(k_i,i)}, \mathbf{L}_{task}^{(k_{i-1},i-1)})$.

By discretizing the search space, the original problem is converted to a combinatorial one, which can be solved by using conventional way, e.g. Dijkstra. However, this straightforward approach is extremely time-consuming and can be hardly accepted for industrial applications. For example, it takes over 20 hours to find a desired solution in a relatively simple case (two-axis robot and one-axis positioner), where the search space is built for 100 task points and the discretization step 1° (processor Intel® i5 2.67 GHz)[12]. Besides, known methods are not able to take into account the acceleration constraints that are necessary here. For these reasons, a problem-oriented algorithm taking into account the particularities of the graph based search space is proposed in this paper.

The developed algorithm is based on the dynamic programming principle, aiming at finding the shortest path from $\{\mathbf{L}_{task}^{(k_1,1)}, \forall k_1\}$ to the current $\{\mathbf{L}_{task}^{(k_i,i)}, \forall k_i\}$. The length of this shortest path is denoted as $d_{k,i}$. Then, the shortest path for the locations corresponding to the next $\{\mathbf{L}_{task}^{(k,i+1)}, \forall k\}$ can be obtained by combining the optimal solutions for the previous column $\{\mathbf{L}_{task}^{(k',i)}, \forall k'\}$ and the distances between the task locations with the indices i and i+1,

$$d_{k,i+1} = \min_{k'}\{d_{k',i} + dist(\mathbf{L}_{task}^{(k,i+1)}, \mathbf{L}_{task}^{(k',i)})\}; \tag{6}$$

This formula is applied sequentially from the second column of the task graph to the last one, and the desired optimal path can be obtained after selection of the minimum length $d_{k,i+1}$ corresponding to the final column. Therefore, the desired path is described by the recorded indices $\{k_1, k_2, ...k_n\}$ .This proposed algorithm is rather time-efficient since it takes about 30 seconds [12]to find the optimal solution for the above mentioned example.

4. Tuning of trajectories generation algorithm

For the proposed methodology, the discretization step for the redundant variable is a key parameter, which has a big influence on the algorithm efficiency. An unsuitable discretization step may lead eitherto a bad solution or high computational time. For this reason, a new strategy for the determination of the discretization step is proposed thereafter to tune the optimization algorithm.

4.1. Influence of the discretization step

Let us consider a simple case study that deals with a three-axis planar robotic system executing a straight-line-task (see Figure 3). For this problem, the fiber placement path isuniformly discretized into 40 segments.Relevant optimization results are presented in Table 1. It is clear that here (as well as in other cases) smaller discretization step should provide better results but there exists a reasonable lower boundrelated to an acceptable computing time.

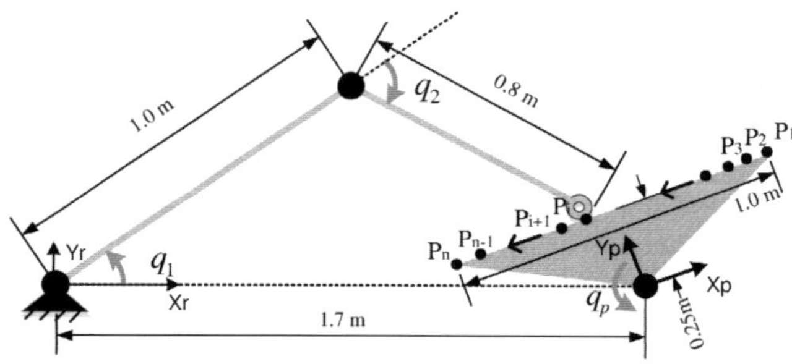

Figure 3. Three-axis planar robotic system andstraight-line-task

Table1. Optimization results and computing time for different discretization steps

	$\Delta q_P = 2°$	$\Delta q_P = 1°$	$\Delta q_P = 0.75°$	$\Delta q_P = 0.5°$	$\Delta q_P = 0.25°$	$\Delta q_P = 0.1°$
Robot processing time (without acc-constraint)	T=1.90s (~38s comp.)	T=1.84s (~2m comp.)	T=1.54s (~4m comp.)	T=1.30s (~9m comp.)	T=1.29s (~47m comp.)	T=1.29s (~5h comp.)
Robot processing time (with acc-constraint)	T=1.90s (~67s comp.)	T=2.11s (~4m comp.)	T=1.60s (~8m comp.)	T=1.30s (~17m comp.)	T=1.29s (~1.2h comp.)	T=1.29s (~9h comp.)

To estimate the reasonable discretization step for the considered fiber placement problem, let us analyze Table 1 in more details. From Table 1, the discretization steps $\Delta q_P \in \{2°, 1°, 0.75°\}$ are not acceptable because they lead toa robot processing time 20-50% higher than the optimal one. Moreover, in the case of $\Delta q_P = 2°$, the optimization algorithm generates a badsolution that does not take advantage of the positioner motion capabilities (*positioner is locked*, $q_P = const$, $\mathbf{q}_R = var$). The reason for this phenomenon is that the discretization step here is so large that the positioner step-time is alwayshigher than the robot moving time between the subsequent task points.

Another interesting phenomenon can be observed for slightly smaller discretization steps, where the algorithmmay produce non-smooth intermittentrotation of the positioner(*start-stop motion*: $q_P = const, ..., (q_P + \Delta q_P), ..., q_P = const$). For example, in the case of $\Delta q_P = 1°$ (without acceleration constraints), the optimization algorithm generates solution that includes only several steps where the positioner is not locked.

In addition, it is noteworthy that in the case of with acceleration constraints, the discretization step reduction from 2° to 1° leads to even worse solution, where the robot processing time is about 10% higher. This phenomenon can be explained by heuristic integration of the acceleration constraints into the optimization algorithm, which may slightly violate the dynamic programming principle.Nevertheless, further reduction of Δq_P allows to restore the expected algorithm behavior. Hence, to apply the developed technique in practice, users need some simple "rules of thumb" that allows setting an initial value of Δq_P. Then, the optimization algorithm can be applied several times (sequentially decreasing Δq_P) until the objective functionconvergence. To reduce computing time in the case of small Δq_P, some local optimization techniques have been also developed by the authors.

4.2. Initial tuning of the optimization algorithm

To find a reasonable initial value of the discretization step, let us investigate in details robot and positioner motions between two sequential task locations. It is clear that for smooth positioner motions, it is required that corresponding increments of the coordinate q_P should include at least one discretization step Δq_P.

To find the maximum value of Δq_P, let us denote $\Delta\theta$ as the increment of q_P for the movement between two adjacent task locations (P_i-P_{i+1}) and Δs as the length of the path segment. It is clear that Δs can be alsotreated as the arc length between P_i and P_{i+1} around the positioner joint axis. Let us assume that the distance from a path point to the rotational axis is r, and r_{max} represents the furthest task locationwith respect to the positioner axis.To avoid undesired intermittent positioner rotations, the following constraint $\Delta s > r_{\max} \cdot \Delta\theta$ should be verified, sincethe positioner velocity is usually smaller than the velocity of the robot.The latter inequality can be rewritten in terms of the robot/positioner motion time as $(\Delta s - r_{\max} \cdot \Delta\theta)/v_R^{\max} = \Delta\theta/\dot{q}_P^{\max}$, which can be further $\Delta q_P \le \Delta\theta$ where number of the positioner steps is no less one. Hence, the initial value of Δq_P should be at least equal to $\dot{q}_P^{\max} \cdot \Delta s/(v_R^{\max} + \dot{q}_P^{\max} \cdot r_{\max})$ in order to provide acceptable motions of the robot and positioner. For instance, for the previous case study, this expression gives the discretization step about 0.5° that allows to generate trajectories that are very close to the optimal ones, namely, the robot processing time is only 1% higher than the minimum value.

5. Conclusions

This paper contributes to optimization of robot/positioner motions in redundant robotic systems for the fiber placement process.It proposes a newstrategy for the optimization algorithm tuning. The developed technique converts the continuous optimization into a combinatorial one where dynamic programming is applied to find time-optimal motions. The proposed strategy of the optimizationalgorithm tuning allowsessentially decreasing the computing time and generating desired motionssatisfying industrial constraints. Feasibilities and advantages of the presentedtechnique are confirmed by a case study.Future research will focus on its integration into real-life industrial environments (6-axis robot + one-axis positioner).

Acknowledgments

This work has been supported by the China Scholarship Council (Grant N° 201404490018). The authors also acknowledge CETIM for the motivation of this research work.

References

[1] D. Gay, Composite materials: design and applications, CRC press, 2014.
[2] C. Gallet-Hamlyn, Multiple-use robots for composite part manufacturing, JEC composites, (2011) 28-30.
[3] S.T. Peters, Handbook of composites, Springer Science & Business Media, 2013.
[4] K. Kazerounian, A. Nedungadi, Redundancy resolution of serial manipulators based on robot dynamics, Mechanism and machine theory, 23 (1988) 295-303.
[5] S.R. Buss, Introduction to inverse kinematics with jacobian transpose, pseudoinverse and damped least squares methods, IEEE Journal of Robotics and Automation, 17 (2004) 16.
[6] L.B. Gueta, R. Chiba, J. Ota, T. Ueyama, T. Arai, Coordinated motion control of a robot arm and a positioning table with arrangement of multiple goals, Robotics and Automation, 2008. ICRA 2008. IEEE International Conference on, IEEE, 2008, pp. 2252-2258.
[7] L.B. Gueta, R. Chiba, T. Arai, T. Ueyama, J. Ota, Hybrid design for multiple-goal task realization of robot arm with rotating table, Robotics and Automation, 2009. ICRA'09. IEEE International Conference on, IEEE, 2009, pp. 1279-1284.
[8] A. Dolgui, A. Pashkevich, Manipulator motion planning for high-speed robotic laser cutting, International Journal of Production Research, 47 (2009) 5691-5715.
[9] A.P. Pashkevich, A.B. Dolgui, O.A. Chumakov, Multiobjective optimization of robot motion for laser cutting applications, International Journal of Computer Integrated Manufacturing, 17 (2004) 171-183.
[10] B. Zhou, X. Zhang, Z. Meng, X. Dai, Off-line programming system of industrial robot for spraying manufacturing optimization, Proceedings of the 33rd Chinese Control Conference, 2014.
[11] P. Debout, H. Chanal, E. Duc, Tool path smoothing of a redundant machine: Application to Automated Fiber Placement, Computer-Aided Design, 43 (2011) 122-132.
[12] J. Gao, A. Pashkevich, S. Caro, Manipulator Motion Planning in Redundant Robotic System for Fiber Placement Process, in: P. Wenger, P. Flores (Eds.) New Trends in Mechanism and Machine Science: Theory and Industrial Applications, Springer International Publishing, Cham, 2017, pp. 243-252.

IOP Conf. Series: Materials Science and Engineering **212** (2017) 012010 doi:10.1088/1757-899X/212/1/012010

Study of Dynamic Characteristics for Hydraulic System on 300MN Die-forging Press

Guoqiang CHEN[1, 2,*], Jianping TAN[1]

[1] College of Mechanical and Electrical Engineering, Central South University, Changsha 410083, China;
[2]College of Mechanical Engineering, Hunan Institute of Engineering, Xiangtan 411104, China

*chengq1979@163.com

Abstract. The faults such as seal breakdown and pressure sensor damage occur in 300MN Die-forging press frequently. First, the fault phenomenon and harm of the hydraulic system was compiled statistics, the theoretical analysis of the hydraulic impact of hydraulic system are carried out based on the momentum theorem; Then, the co-simulation model of hydraulic system was established by AMESim and Simulink software and the correctness was verified. Finally, the dynamic characteristics of hydraulic system for the key working condition "forging stroke changing to mold collision" was analyzed, the influences rules of system parameters such as the leak gap of valve, diameter of water way pipeline, emulsion temperature and air contain act on hydraulic system are obtained. This conclusions have a theoretical guiding significance to the improvement and maintains of high pressure and large flow hydraulic system.

1. Introduction

The hydraulic system of 300MN Die-forging press has the characteristics of high pressure and large flow rate, the faults such as seal breakdown and pressure sensor damage occur frequently during using. In order to realize the performance indicators, the stable dynamic characteristics were needed especially[1].In general, the dynamic characteristics of the hydraulic system is mainly manifested in the amplitude of pressure wave. The numerical simulation method has become an effective methods to study the dynamic characteristics of hydraulic system[2-5]. In this paper, the hydraulic system of 300MN die forging press was taken as research object, the hydraulic impact mechanism and dynamic characteristics of hydraulic system with high pressure and large flow were systematic studied based on theoretical analysis and numerical simulation.

2. The structure and hydraulic system of 300MN die forging hydraulic press

The hydraulic system of 300MN die forging hydraulic press includes lift cylinder, balance cylinder, working cylinder, pump station, main distributor pipeline, filling valve, filling tank and other matching components. The lift cylinder, balance cylinder and working cylinder of hydraulic press were connected with the machine body, the working cylinder is fixed on the upper beam, the lifting cylinder and balance cylinder is fixed on the lower beam, each plunger of all cylinders connected on

IOP Conf. Series: Materials Science and Engineering **212** (2017) 012010 doi:10.1088/1757-899X/212/1/012010

moving beam together. The normal working pressure is 320bar 300MN die forging hydraulic press from the pump station.

The operation of hydraulic system is realized by the main distributor valve, the opening or closing process of valve is controlled by the operating handle which can input angle signal at first, then rotate driving cam to the specified angle, finally, the inlet and drain valve are opening and closing by the push of driving cam, the valve should be in accordance with the laws of coordination in order to achieve the conversion of four kinds of working state: stop, virtual travel, forging and lift, the operation handle angle signal can be converted to the sequence order and the height of valve opening signal, the valve opening and closing angle and the relationship with the operation handle as shown in Figure 1, where 1-8 is the number of valve:1,3 - inlet valve of working cylinder, 2- the drain valve of working cylinder, 5- inlet valve of balance cylinder, 4- drain valve of balance cylinder, 7- inlet valve of lift cylinder, 8- drain valve of lift cylinder.

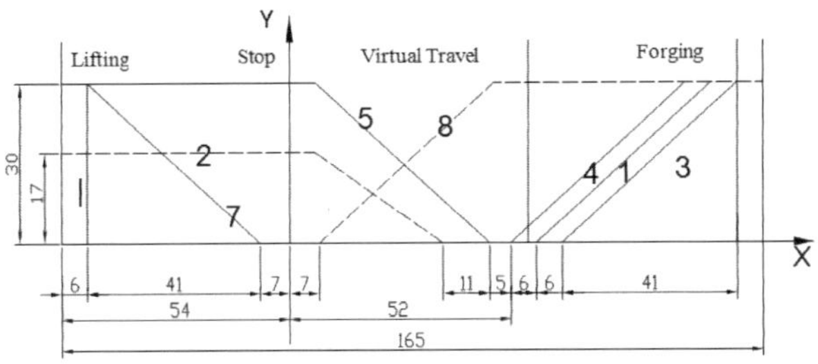

Figure 1. The opening control chart of valve core for main distributor

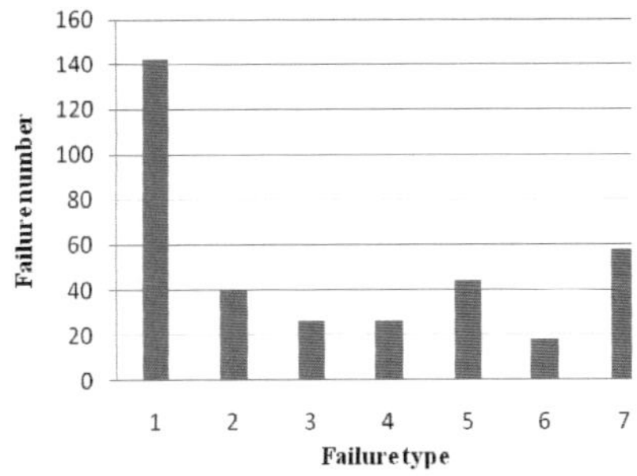

Figure 2. The failure statistics of 300MN die forging press

In Figure1,the X-axis represents the handle rotation angle, the Y-axis is the opening height of the related valve. The "stop" indicating the position of the cam angle 0 degrees, the cam angle value is negative during top to lifting, cam angle is positive during stop to forging.

3. Hydraulic impact mechanism analysis of waterway system

3.1 Hydraulic impact of waterway system on 300MN die forging hydraulic press

The hydraulic impact, which has great impact on the performance of stability and reliability of the hydraulic system, widely exist in various types of hydraulic systems, and is a problem often encountered in the working process of the system. Studies show that the high pressure and large flow of hydraulic pressure is the main cause of seal failure in the system[6], and will lead to vibration noise ,loose connections and other phenomena, even the occurrence of pipeline rupture, hydraulic components and measuring instrument damage.

According to the failure statistics of mechanical and hydraulic part on 300MN die forging hydraulic press during 10th Nov.,2008 to 24th Nov.,2009, the total of 354 times of failures and its distribution are shown in Figure2, where, 1-Valve seal leakage, 2-Valve opening failure, 3-Pipeline loose, 4-plate loose, 5-flange leakage, 6- Hydraulic cylinder leakage, 7 - other faults; As we can see from Figure2 that the seal failure, pipeline loose flange leakage fault accounted for the proportion of the failure of the entire system is great, seriously affect the normal production equipment.

3.2 Hydraulic impact analysis based on momentum theorem

There two reasons for the hydraulic impact of hydraulic system, the first one, the closing of valve can lead the high pressure flow liquid in pipeline stopped quickly, the inertial would cause the hydraulic impact inevitable; the second one, with the status changed, the huge inertial of moving parts such as beam can cause the hydraulic impact obviously, the hydraulic impact and dynamic characteristics is more larger especially when the state transiting quickly.

In the whole process, the hydraulic system will follow the law of conservation of momentum, so we can complete the theoretical analysis and accurate calculation of hydraulic impact based on momentum theorem. The working cycle of 300MN die forging hydraulic press comprises stop, virtual travel, and forging and lifting, the scene shows that when hydraulic press in “virtual travel to lifting” and “forging stroke changing to mold collision”, the switching process appear obvious vibration. In here, the state switch of “virtual travel to lifting” was taken as the object to analyze the hydraulic impact in theoretically.

3.2.1 The Hydraulic impact of lift pipeline when the valve closing Δp_1

Set ρ as the density of liquid, υ is the liquid velocity, the valve flow rate is zero after the closure of valve, follow the momentum equation:

$$\Delta p_1 = \rho c \upsilon \quad (1)$$

The parameter c is not only related to the bulk modulus K, but also to the elastic modulus of pipeline material E , the inner diameter d and the thickness of the pipeline δ, which can be calculated by :

$$c = \sqrt{\frac{K}{\rho}} / \sqrt{1 + \frac{Kd}{E\delta}} \quad (2)$$

Here, ρ—fluid density ,1017.9kg/m^3 ; K--Liquid bulk modulus, 2.06×109N/m;

E—Elastic modulus of pipeline material, 2.2×109N/m; The pipeline diameter of lifting cylinder d=112mm; the wall thickness δ=26mm; the beam maximum speed: 150mm/s; $S_{cylinder}$= 0.7235m^2, the flow rate of the valve port υ:

$\upsilon = V_{beam} \cdot S_{cylinder} / S_{pipeline}$=0.15*0.7235/0.009847=11.2 (m/s)

Follow the formula (2), parameter c can be calculated as c=1270 (m/s).

Based on the momentum theorem formula (1), Δp_1 = 14.47Mpa.

3.2.2 The hydraulic impact caused by moving parts during braking Δp_2

Set the deceleration time of moving parts which total mass $\sum m$ for Δt , Δv as the speed decreases value, the effective area of hydraulic cylinder as A, according to the theorem of momentum:

$$\Delta p_2 = \frac{\sum m \Delta v}{A \Delta t} \quad (3)$$

In 300MN die forging hydraulic press, the moving parts includes a movable beam and the lift cylinder piston, which the total mass of movable beam is 2.3×10^6Kg, the piston mass is too light compared with the beam, so it can be ignored, the maximum velocity of movable beam air travel is 0.15m/s, piston total area of lifting cylinder is 0.7235m^2, take the above parameters into formula (3), the hydraulic impact caused by movable beam in the fall time Δt can be obtained:

$$\Delta p_2 = 0.467 / \Delta t \quad (4)$$

3.2.3 The total hydraulic impact during virtual travel to lifting Δp

For the movable beam falling from the virtual travel to a sudden stop, the hydraulic impact generated by follow formula(5):

$$\Delta p = \Delta p_1 + \Delta p_2 = 14.47 + {0.467}/{\Delta t} \quad (5)$$

Set p_0 as the initial internal pressure of pipeline, in the virtual travel, the gravity of the movable beam is balanced by the balance cylinder and the lifting cylinder, so it can be regarded as the initial pressure of the lifting loop system at this time is 9.68MPa, so the total pressure in pipeline is follow formula:

$$p_{max} = p_0 + \Delta p = 24.15 + {0.467}/{\Delta t} \quad (6)$$

4.The dynamic characteristics analysis of hydraulic system

IOP Conf. Series: Materials Science and Engineering **212** (2017) 012010 doi:10.1088/1757-899X/212/1/012010

The AMESim software has the characteristics of simple and intuitive in the hydraulic simulation modeling work, and the Simulink software has Superior performance in data processing, so, this paper adopts the co-simulation of AMESim and Simulink software to analyze the hydraulic impact and dynamic characteristics of the hydraulic system.

4.1 Simulation model of hydraulic system

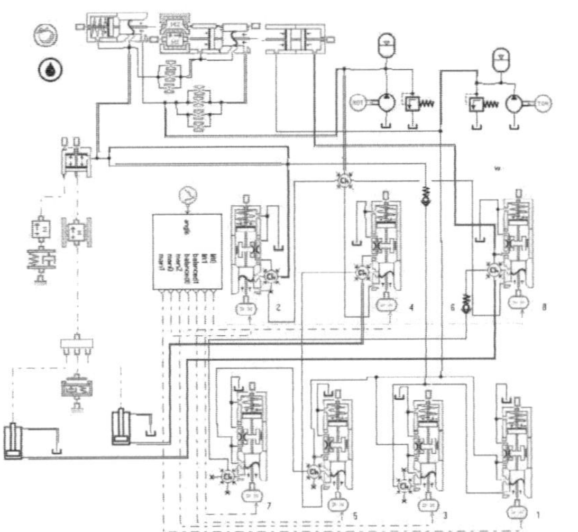

Figure3. AMESim model of hydraulic system

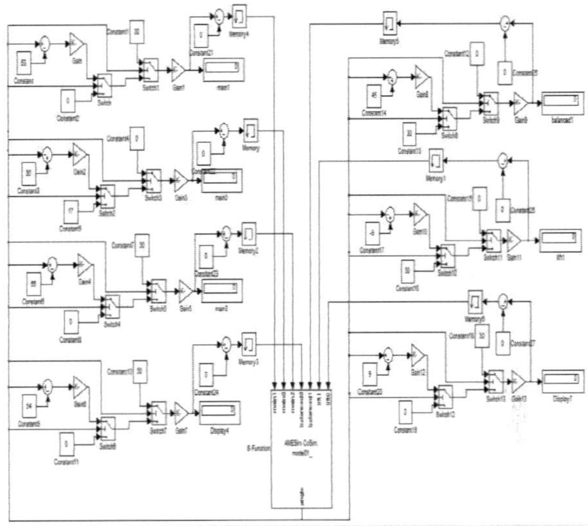

Figure4. Simulink model of opening control function

Because there are 16 hydraulic cylinder in this hydraulic system, it cause the system model becomes complicated and low computational efficiency, thus we need simplify the hydraulic cylinder according to the cross-sectional area of the equivalent principle of system, hydraulic system simulation model is shown in Figure 3, the valve opening control function model is established in the simulink software as shown in Figure 4

4.2 Theoretical verification of simulation model

According to the characteristics of the numerical simulation, the veracity and reliability of system model is uncertain. The Hydraulic press in the process of "virtual travel to lifting" was taken as object, the hydraulic impact of theoretical analysis and calculation based on the theorem of momentum compared with the results of simulation, the numerical simulation model can be verified.

Take the cam angle conversion time (valve action time) for Δt =0.1s, using the formula (6) can be calculated the pressure of the pipeline at this time is 28.82MPa, that is, 288.2Bar; Then, under same conditions and the valve closed within 0.1second begin at the 2th seconds, the pressure curve of hydraulic system based on the numerical simulation model shown in Figure 5:

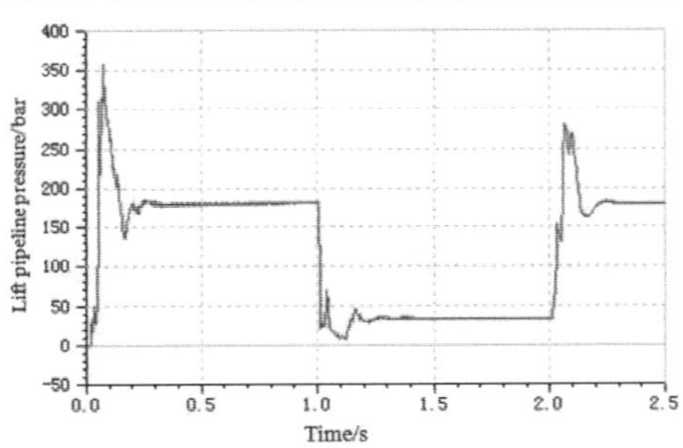

Figure 5. the pressure curve of virtual travel to lifting process

Comparing the theoretical calculation (288.2 bar) and the simulation results (278.4 bar) shows that the error between the simulation and theoretical calculation is only 9.8bar, and the error rate of 3.4% ,the error be in the permitted range, so we can determined that the simulation model is Accurate and reliable.

4.3 The simulation analysis for dynamic characteristics of hydraulic system

The process of "forging to mold collision" is a dangerous action of the hydraulic press work, while, there are high pressure in the work cylinder and valve have no action, the change of forging deformation resistance leads to mold collision, the movable beam stopped suddenly. The influence law of leakage gap, air content, emulsion temperature, and the diameter of pipeline for the dynamic characteristics under this conditions have been founded based on simulation.

As the reciprocating motion of valve stem, the gap is inevitable exist in the valve, cause the internal leaks especially serious under the background of high pressure. The pressure peak of working cylinder as shown in Figure 6, it shows that the leakage of the valve body effect the cylinder pressure very obviously, especially when the leakage gap reached more than 1.5mm.

Set the emulsion air content of 0.1% to 0.1% in 11 sets of data for simulation analysis, the working cylinder pressure as shown in Figure 7.

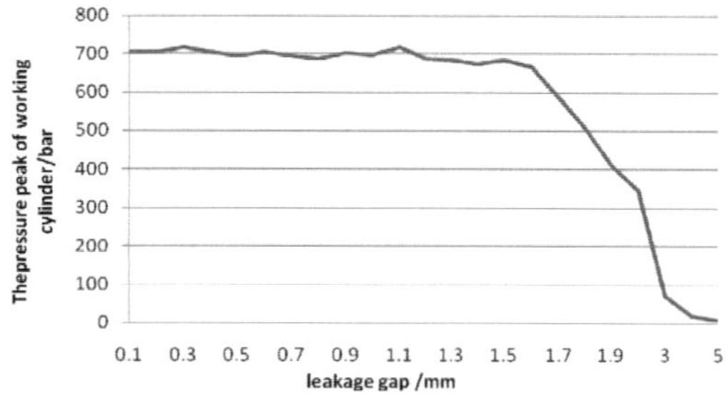

Figure 6. The pressure curve with leakage gap

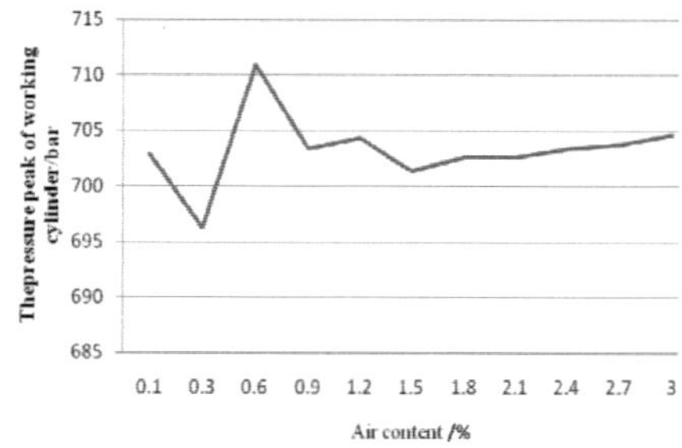

Figure 7. The pressure curve with air content

The figure 7 shows that the air content is between 0.1% to 3%, the impact on the work in cylinder pressure amplitude is not very big, when the air content reach to 1.5%, the effect tends to stability and gradually increase.

Due to the existence of energy conversion and consumption in the work process, the emulsion medium inevitably will be warming, based on the analysis of emulsion temperature is 5℃-70℃ in 14 groups of data, obtained the pressure curve as shown in Figure 8:

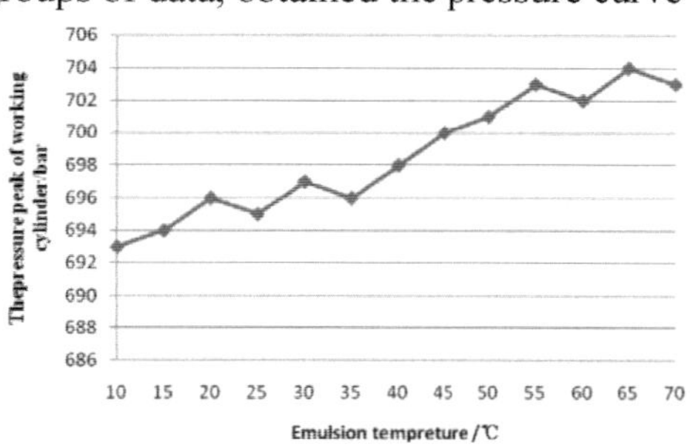

Figure 8. The pressure curve with temperature

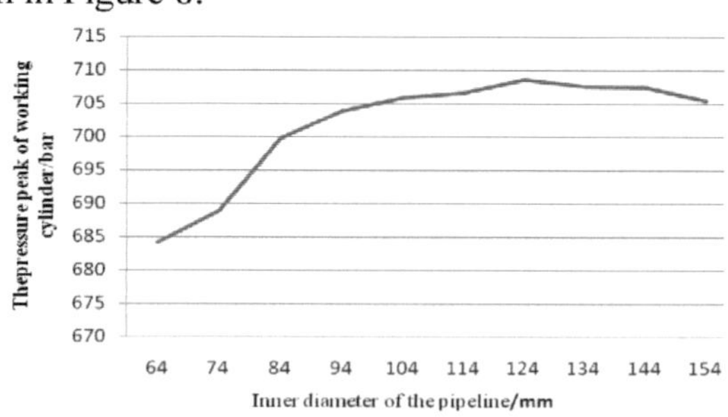

Figure 9. The pressure curve with pipeline diameter

The figure 8 shows that the pressure peak in working cylinder is increasing with the temperature's rising, And similar to the linear relationship. In the scene, it is necessary to keep the emulsion temperature under 40℃;

Set the inner diameter of working cylinder pipeline for 9 sets of data in the 64 mm-154mm, and then obtained the pressure curve after simulation and analysis, as shown in Figure 9, it shows that the

IOP Conf. Series: Materials Science and Engineering **212** (2017) 012010 doi:10.1088/1757-899X/212/1/012010

cylinder pipe diameter is smaller, the impact of the pressure peak of the cylinder is smaller, and more than 124mm, the effect for the pressure peak is tend to stability.

5. Conclusions

The fault phenomenon and hydraulic impact mechanism of 300MN die-forging hydraulic press were analyzed, the co-simulation model of hydraulic system was established based on AMESim and Simulink software, the dynamic characteristics of hydraulic system during “forging stroke changing to mold collision” have been founded, studies show that this method is effective and credible. Then the influence law of leakage gap, air content, emulsion temperature, and the diameter of pipeline for the dynamic characteristics under have been founded based on simulation. This results can be applied to the design and maintenance of hydraulic system with high pressure large flow.

Acknowledgments

This work was partially supported by National science and technology major project (No.2009ZX04005-031); National Program on Key Basic Research Project (973 Program) under Grant No. 2014CB049405

Reference

[1] Feng shuangchang. The Modeling and Simulation of dynamic characteristics for hydraulic system of excavator working device[D]. Guiyang,Guizhou university,2007;

[2] Manfen Han, Yantao Song, Wenlong Zhao. Simulation and Optimization of Synchronization Control System for CFETR Water Hydraulic Manipulator Based on AMEsim[J]. Journal of Fusion Energy,June 2015,Volume 34,Issue3,Pages 566–570;

[3] H.P. Wang, D. Zheng, Y. Tian. High pressure common rail injection system modeling and control[J].ISA Transactions ,Volume 63, July 2016, Pages 265–273;

[4] Asgat Gimadiev, Pavel Greshnyakov, Dmitry Stadnik. Development of Mathematical Model and Analysis of Characteristics of Gas Dampener for Power Plant Fuel Main Line[J].Procedia Engineering, Volume 106, 2015, Pages 53-61;

[5] LIU Jiang-ming, TAN Jian-ping, CHEN Hui. Hydraulic impact Simulation on the progress of descending changing to lifting stroke in 300MN hydraulic press based on AMESim and Simulink[J]. Hydraulics Pneumatics & Seals, Volume 7, 2012, Pages 24-27;

[6] CHEN Guoqiang, TAN Jianping, CHEN Hui. Failure mechanism of U-ring seal in water valve with high pressure and large flow capacity[J]. Journal of Central South University (Science and Technology), Volume 44,2013, Pages 44:1-7;

IOP Conf. Series: Materials Science and Engineering **212** (2017) 012011 doi:10.1088/1757-899X/212/1/012011

Pneumatic Muscle Actuated Rehabilitation Equipment of the Upper Limb Joints

Prof.dr.eng.habil. Andrea Deaconescu

Department of Industrial Engineering and Management, Transilvania UniversityBrasov, Romania

e-mail: deacon@unitbv.ro

Abstract. Rehabilitation equipment of the upper limb joints holds a key role in passive physical therapy. Within this framework, the paper presents two such pieces of equipment developed for the rehabilitation of elbow and of wrist and knuckles, respectively. The presented and discussed equipment is actuated by pneumatic muscles, its benefits being a low cost, simple and robust construction, as well as short response time to commands.

1. Introduction

The mobility of the limbs is of essence for personal physical independence. For persons suffering from post-traumatic affections of the upper limbs, the lack of mobility creates barriers in carrying out day-to-day tasks, both personal and professional. In the wake of accidents and surgical treatment, patients are often subjected to a period of immobilisation of the limbs. Prolonged immobilisation cause muscle hypertrophy, dysfunctions of the circulatory apparatus and bone demineralization. For this reason a follow-up rehabilitation programme is always necessary.

The main objective of post-traumatic recovery is characterised by:

- maintaining the normal motion amplitude of the healthy segments that were not affected by trauma or surgical interventions;
- recovery of parts affected by trauma (muscle, bone, ligament, tendon tissue, etc.);
- regaining of the initial functions of the body, the affected segments, etc.

The most important method used for regaining the diminished functions and increasing joint mobility is physical therapy, that includes techniques focused on mobilising the targeted segment (active or static mobilisation techniques), as well as techniques that do not entail its motion (immobilisation and postural therapy). Physical therapy is efficient in the case of both minor and major trauma.

Physical therapy, together with kinetic prophylaxis are components of kinesiology, an interdisciplinary biological science that studies the movements of the human body, the functional elements that contribute to such movement and to the compensation modalities of reversible, partially reversible and irreversible disturbances [1].

On kinesiology nearly 5000 year old written information was discovered [2]. Thus Chinese documents dated 4700 years ago describe therapeutic postures and motions meant to ease pain. Also the Indian Vedas feature exercises recommended in rheumatism and other conditions. The true foundations of kinesiology, however, were laid by its founder Aristoteles in ancient Greece, followed by Hippocrates who observed that movement helps regaining muscle force (Fig. 1), while immobilisation leads to muscle atrophy.

Figure 1. Mobilisation of wrist and knuckles by traction (Hippocrates).

At present, in wrist and knuckle physical therapy, their mobilisation is achieved gently, the patient regaining the natural movements in a conscious manner. The alternative for patients unable to carry out the imposed motions consciously is passive mobilisation.

Continuous Passive Motion (CPM) is a therapeutic procedure that consists in applying a series of movements to the affected joint, without, however, entailing a contraction of the patient's muscles. This is achieved with the help of a physical therapist or by means of specially designed equipment.

Rehabilitation equipment plays an important role in passive physical therapy. Although passive physical therapy can be performed by the physical therapist without the help if such equipment, they are nevertheless deployed for moving the affected limb, thus easing the therapist's work.

The development of new mechanical rehabilitation methods of the upper limb joints has inevitably led to the conceiving of specific recovery equipment. A number of such pieces of equipment for the recovery of the shoulder, elbow or wrist, respectively, are produced by Rimec, Italy, [3]. Other manufacturers of rehabilitation equipment are Kinetec, France, KLC Services, USA and Ormed Medical Technology, India [4 - 6].

The conducted research has revealed that most rehabilitation equipment of the upper limb joints currently available on the marketplace is actuated by electric motors and has relatively high prices, of thousands of euros. For this reason at the National Research and Training Centre for Fluidic Drives and Automation at Transilvania University of Braşov, Romania such equipment – but pneumatically driven, were designed and developed. This paper presents two models of rehabilitation equipment that achieve continuous passive motion of the elbow, the wrist and the knuckles, respectively. These pieces of equipment are actuated by pneumatic muscles, the energy source generating force and motion being compressed air. The main advantages of the actuation system with pneumatic muscles are the low cost, simple and robust construction and swift reaction to commands. Due to the simple construction and reduced cost, the proposed equipment can be used in medical units by qualified personnel, as well as outside such entities by non-specialist personnel.

2. Wrist rehabilitation equipment

2.1. Rehabilitation motions and their limits

Pneumatically actuated wrist rehabilitation equipment is the object of several patents, some of which are described in [7] and [8].

The construction of such rehabilitation equipment starts from identifying the motions required for recovery, and of their respective amplitudes. The movements of the hand (the fist) are extremely complex and are facilitated by several joints, like the radiocarpal, the intercarpal or metacarpal ones. Flexion and extension of the fist, respectively, are carried out by the metacarpal joint as follows:

- Flexion of the fist or flexion of the palm – is obtained by rotating the palm in volar direction. The zero position is obtained when the forearm is flexed at 90º and in pronation (Fig. 2). The maximum rotation angle is of 90º according to [9].
- Extension of the fist or dorsiflexion – is obtained by rotating the palm in dorsal direction. The maximum rotation angle is of 70º according to [9] (Fig. 2).

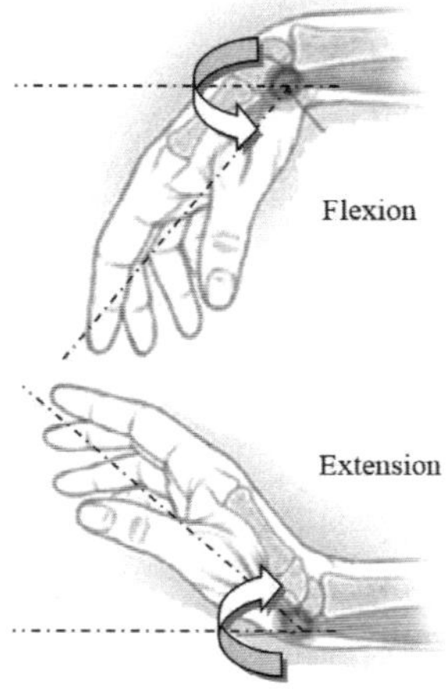

Figure 2. Flexion/extension movements of the fist.

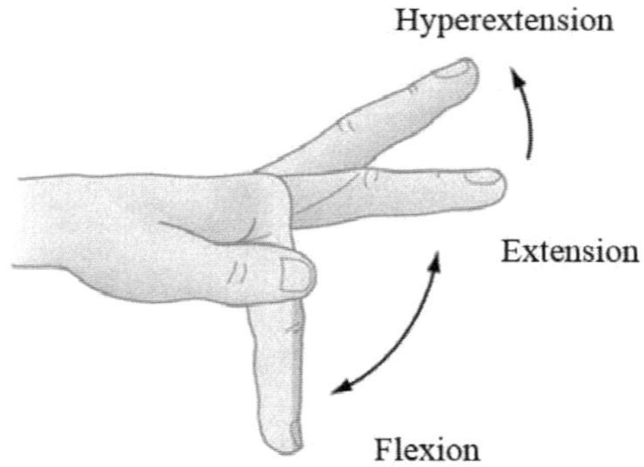

Figure 3. Flexion/extension movements of the fingers.

In the case of the knuckles, flexion – extension takes place within the following ranges:

- The flexion of the fingers increases from finger II to finger V from 90º to 100º, when fingers II, III, IV and V are flexed simultaneously (Fig. 3), according to [9].
- The extension of the fingers (Fig. 3) varies from individual to individual and can range from 0º to 90º (hyperextension), according to [9].

Fig. 4 features the proximal and distal inter-phalanx joints of phalanges II , III, IV şi V. Proximal flexion is the movement of the middle phalanx towards the inner palm, towards the proximal phalanx, until the limit of the movement is reached. The amplitude of this movement is of 100º, according to [9] or of 120º, according to [10]. Flexion has smaller values for fingers I and II, ad greater values for fingers IV and V.

Distal flexion is the movement of the distal phalanx towards the inner palm, towards the middle phalanx, until the limit of the movement is reached. In order to obtain the maximum value, the proximal joint needs flexing. The maximum value of this joint's flexion does not exceed 90º, according to [9] or 80º, according to [10].

Extension is possible only for the distal joints, the maximum amplitude being 20º, according to [9], it being present only in some individuals.

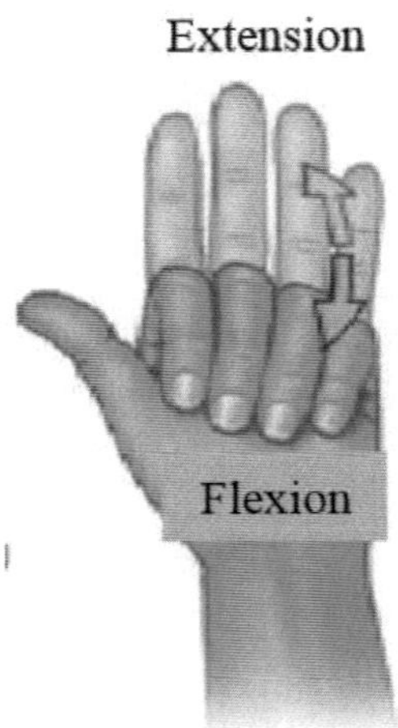

Figure 4. Proximal and distal flexion.

2.2. The constructive solution

Knowing the movements of the wrist and fingers, as well as their amplitudes, a rehabilitation device was conceived the motions of that are achieved by original bio-inspired systems based on the Fin Ray effect, specific for fish fins. There are similarities between the movements of the palm and fingers and the fishtail movement, as the flexion-extension of the palm is similar to the oscillatoy motion of the tail fin of some fish (Fig. 5). The bone structure of the hand and of the fishtail, respectively, is set into motion by muscle groups located on both sides of the skeleton and operating in countertime.

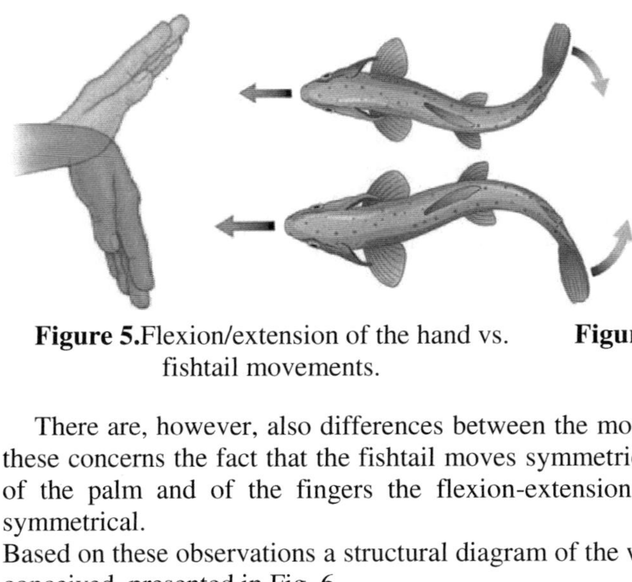

Figure 5. Flexion/extension of the hand vs. fishtail movements.

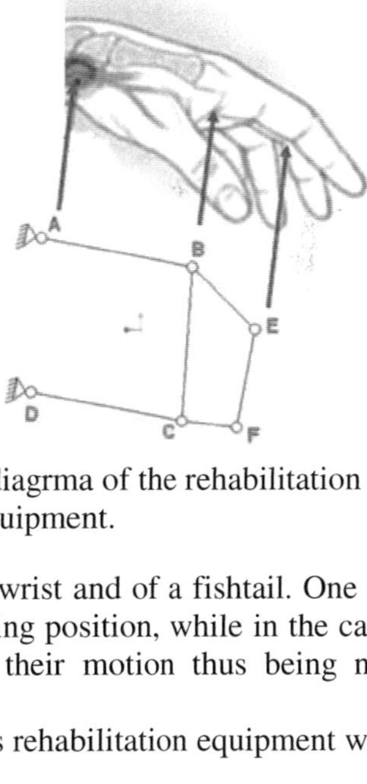

Figure 6. Structural diagrma of the rehabilitation equipment.

There are, however, also differences between the movements of the wrist and of a fishtail. One of these concerns the fact that the fishtail moves symmetrically by its resting position, while in the case of the palm and of the fingers the flexion-extension angles differ, their motion thus being not symmetrical.

Based on these observations a structural diagram of the wrist and fingers rehabilitation equipment was conceived, presented in Fig. 6.

The assembly that supports and mobilises the palm consists of two bar mechanisms connected by torsion springs, forming a Fin ray type structure. The length of segment AB is given by the distance between the fist joint (A) and the metacarpal-phalanx joints (B), and segment BE represents the length between the metacarpal-phalanx joints (B) and the proximal joints (E).

This bar system is set into motion at joint A, by a gear -rack mechanism, the latter's movement being imposed by a pneumatic muscle (Fig. 7).

IOP Conf. Series: Materials Science and Engineering **212** (2017) 012011 doi:10.1088/1757-899X/212/1/012011

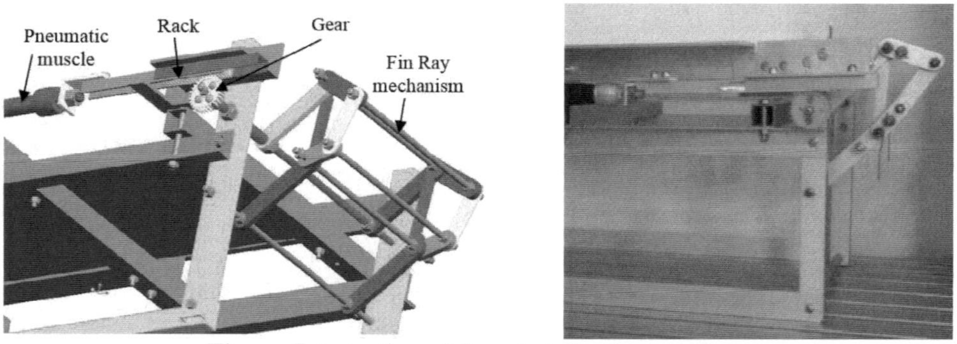

Figure 7. Actuation of the rehabilitation system.

The resting position of the system corresponds to the flexion of the hand and corresponds to the situation in that the pneumatic muscle is not fed air. Feeding air into the muscle causes the retraction of the rack and implicitly, the rotation of the bar system such as to achieve the extension of the hand.

Fig. 8. presents the positions of the hand on the bar system in the case of maximum flexion and extension of the hand.

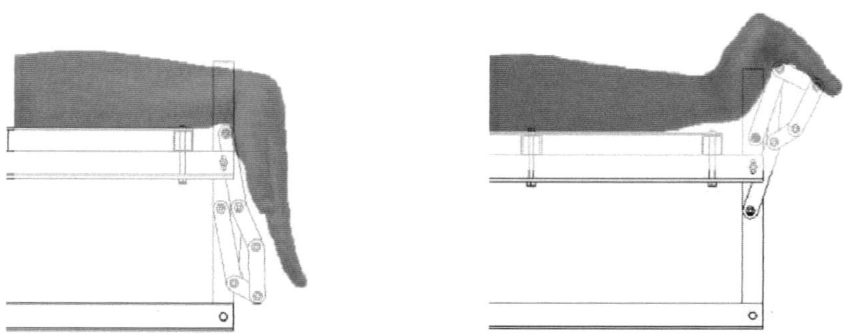

Figure 8. Flexion and extension of the hand.

Hand flexion and extension at various angles is obtained by corresponding commands of the pneumatic muscle stroke. A higher feeding pressure of the pneumatic muscle causes a greater rotation of the wrist in the direction of extension.

Table I presents a synthesis of the maximum angles achieved by the Fin Ray mechanism and of those found in literature.

Table1. Maximum flexion/extension angles.

	Flexion of the fist	Extension of the fist	Flexion of the fingers	Extension of the fingers
Found in literature	70°	80°	90°-100°	0°- 90°
Achieved by the mechanism	70°	80°	102°	9.67°

In addition to the fact that the equipment satisfies the functional requirements for the rehabilitation of the hand joints, its manufacturing cost is low, of only a couple of hundred euros.

Another benefit of this equipment follows from the fact that the other types of equipment available on the marketplace are specialised on either the rotation of the wrist or on the rotation of the fingers The proposed equipment, on the other hand, allows the simultaneous mobilisation of both hand and fingers.

3. Rehabilitation equipment of the elbow

3.1. Rehabilitation motions and their limits
The pneumatic actuation of elbow rehabilitation equipment has been the object of several research projects and patents have been granted to some Chinese universities [11 - 14].

In this case too, a prerequisite for constructing such rehabilitation equipment is identifying the movements required for recovery and their angular amplitudes.

The elbow is a complex and mobile joint, formed by the inferior part of the humerus and the superior part of the radius and cubitus (ulna). The elbow is a joint that ensures two degrees of mobility and allows flexion-extension and pronation-supination. Characteristic for flexion-extension is the humeroulnar, and for pronation-supination the radioulnar articulation.

Flexion, exemplified in Fig. 9a, is obtained by moving the forearm towards the arm, the maximum rotation angle being of 150°. Extension is obtained by moving the forearm away from the arm, its values ranging from 5° to 10° (hyperextension).

Pronosupination consists in the rotation of the forearm by its longitudinal axis (Fig. 9b), the amplitude of this rotation being of ±90°.

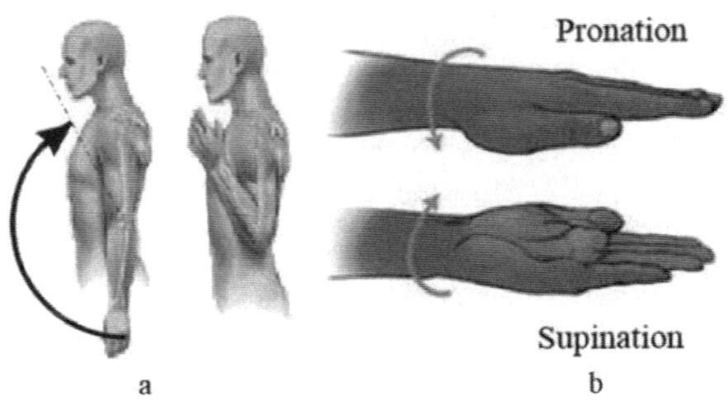

Figure 9. Elbow movements.

3.2. The constructive solution
The proposed rehabilitation equipment has two degrees of mobility, materialised by rotations conducted by two perpendicular axes. Each rotation is obtained by means of a pair of pneumatic muscles. Figures 10 and 11 show the diagrams of principle underlying these two movements.

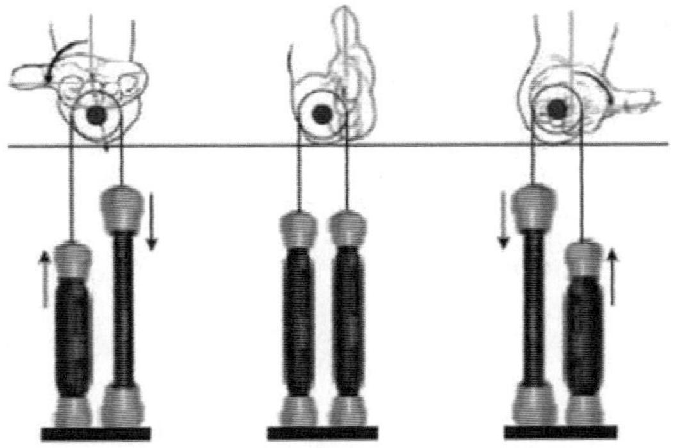

Figure 10. Generating pronosupination.

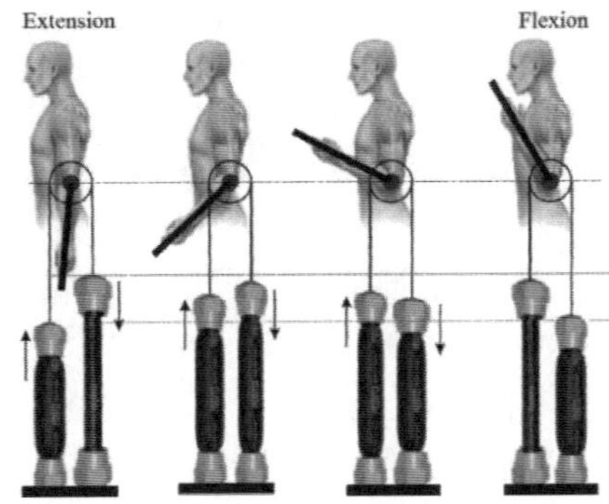

Figure 11. Generating flexion/extension.

It can be noticed that the two rotations are achieved similarly to the operation of human muscles, based on the agonist-antagonist principle, i.e. while a muscle contracts, the other one relaxes. For achieving zero position, both pneumatic muscles are fed the same pressure, upon which, in order to conduct rotation in one or the other direction, one of the muscles is inflated at the same time with the relaxation of the other.

Figure 12 shows the kinematic diagram of the proposed equipment next to a view of a joint.

Figure 12. Diagram of principle of the rehabilitation equipment.

A thread is connected to the free extremities of each pneumatic muscle, stretched over a pulley and fixed in a specially conceived space. This constructive diagram allows the rotation of the pulley in one or the other direction, by the countertime inflation and deflation of the two muscles.

Initially the two pneumatic muscles are at a pressure of $p = 0$ bar, and hang freely from the pulley. In order to generate rotation in one or the other direction, the first step is the simultaneous pre-straining of the two muscles, by feeding them a pressure p_0 equal to half the maximum working pressure. This causes an axial contraction of the two muscles equal to half the maximum contraction generated upon inflating the muscles with a pressure $p = p_{max}$. Thus, when $p_0 = p_{max}/2$, the axial contraction of the two muscles is equal to $\Delta L_{max}/2$, the stroke of the inferior ends of the two pneumatic muscles being limited by a fixed stopper.

When a rotation by a certain angle is desired, one of the muscles will be fed additionally air up to a pressure $p_1 = p_0 + \Delta p$, and the second muscle will relax to a pressure $p_2 = p_0 - \Delta p$. By feeding the two muscles different pressures, their lengths will be modified in relation to their initial state as follows: the muscle inflated to pressure p_1 will shorten to a length of $L_1 = L_0 - \Delta L_{max}/2$, while the second muscle will expand to a length $L_2 = L_0 + \Delta L_{max}/2$. The countertime deformation of the two pneumatic muscles causes the rotation of the articulation, the angular amplitudes of the two axes being of ± 90º.

In the case of the presented equipment, the rehabilitation motions by the two axes are conducted separately, in sequence, and not simultaneously.

4. Conclusion

Upper limb joint rehabilitation equipment actuated by pneumatic muscles becomes a viable alternative, capable of replacing successfully the electrically actuated variants. The developed prototypes prove the high capability and performance of the proposed rehabilitation equipment, the pneumatic muscle being eligible for becoming the adequate actuator of medical recovery systems.

References

[1] S.G. Cioroiu, Elementary Notions of Kinesiology and Massage (in romanian), Brasov, Editura Universității Transilvania, 2003.
[2] T. Sbenghe, Kinesiology – the science of motions (in romanian), Bucuresti, Editura Medicala, 2008.
[3] RIMEC. Products, http://www.rimec.it/en/prodotti_category/passive-mobilization-upper-limb/ Accessed: 15.12.2016.
[4] Kinetec. http://www.kinetec.fr/en/kinetec-selection/cpm-continuous-passive-motion.html Accessed: 15.12.2016.
[5] KLC Services Inc. http://klcflexmate.net/devices.html Accessed: 15.12.2016.
[6] Ormed Medical Technology, http://www.ormedortho.com/artromot_e.htm Accessed: 15.12.2016.
[7] J. Zhang, Y. Liu, Y. Jia, Z. Shen, Device for hand rehabilitation. Patent no. WO2012120393,

2012.

[8] E.J. Koeneman, J.B. Koeneman, D.E. Herring, R.S. Schultz, System and method for neuromuscular reeducation. Patent no. US8214029, 2012.

[9] T. Sbenghe, Prophylactic, therapeutical and recovery kinesiology (in romanian), București, Editura Medicala, 1987.

[10] E.L. Sidenco, Practical guide to joint and muscle assessment in physical therapy (in romanian), București, Editura Fundației „Romania de Maine", 2009.

[11] J. Liu, Z. Yongqiang, Z. Qingtai, Elbow medical rehabilitation device based on pneumatic artificial muscle drive, Patent no. CN203263743, 2013.

[12] J. Liu, H. Xu, H. Lan, J. Ke, X. Cao, Elbow rehabilitation training device, Patent no. CN103976852, 2014.

[13] Y. Dongming, W. Binrui, J. Yinglian, Pneumatic muscle flexible elbow joint device with buffer spring and flexible shaft sleeve, Patent no. CN202071080, 2011.

[14] Y. Canjun, Z. Jiafan, Z. Jie, C. Ying, Flexible exoskeleton elbow joint based on pneumatic power, Patent no. CN200984250, 2007.

IOP Conf. Series: Materials Science and Engineering **212** (2017) 012012 doi:10.1088/1757-899X/212/1/012012

Comparison with Experimental Results of Models and Modelling with Fuzzy Logic of the Effect on Surface Roughness of Cutting Parameters in Machining of Co28Cr6Mowrought Steels

IlhanAsilturk[*], **Mehmet AlperInce**[+]

[*]University of Selcuk, Faculty of Technology, [+]University of Selcuk, Faculty of Engineering

iasiltürk@ selcuk.edu.tr, mehmet4219@ hotmail.com

Abstract. This study includes comparison with experimental results of models and modelling with fuzzy logic of the effect on surface roughness of cutting parameters (rotational speed (*n*), feed rate (*f*), depth of cut (*a*) and tool tip radius (*r*)) in CNC turning of Co28Cr6Mo wrought steels. Fuzzy logic modelswere established that can determine the optimum rotational speed, feed rate, depth of cut and tool tip radius for surface roughness (*R*a) according to the hardness of material and type of cutting tool. In the model created using fuzzy logic, membership functions and foot widths of input parameters and output parameter were utilized. In the rule base, triangular (trim-f) membership functions were selected by the Mamdani approach. The results obtained using this fuzzymodel and experimental results were interpreted and compared with 2dimensional graphics.

1. Introduction

The surface roughnessis important in design.Many researchers have been dealing with the establishment of models in recent years due to the increased number of applied materials and tool materials. These models provide a connection between the cutting parameters and the surface roughness. At the beginning of these models are artificial intelligence techniques.One of these artificial intelligence techniques is fuzzy logic. Professor Zadeh invented fuzzy logic in 1964.
The concept of fuzzy logic came from the idea of grade of membership function, which became the backbone of fuzzy set theory.Fuzzy logic is a basic concept which refers to all theories and technologies that employ fuzzy set [1-2].It allows intermediate values to be defined between evaluations like low/high, slow/fast, small/large etc. Many authors have done some research on surface roughness with fuzzy logic;

A fuzzy based method was proposed by Fei and Jawahir [3] for predicting the achievable levels of surface roughness in finish turning operations. They created the rules from a new method for predicting surface roughness values under cutting conditions.The surface roughness minimization was carried out using fuzzy approach by Gok [4] for nodular cast iron. L_{16} (3^4) Taguchi plan was used to define the experimental run, in which the three input parameters were changed in four levels (Vc=50-100-150-200 m/min; f= 0.05-0.075-0.01-0.125 mm; ap= 0.5-1-1.5-2 mm). The cutting ability of die cast alloy Al-Si-Cu-Fe for four kinds of materials: normal, Bismuth reinforced (Bi),

IOP Conf. Series: Materials Science and Engineering **212** (2017) 012012 doi:10.1088/1757-899X/212/1/012012

Strontiumreinforced (Sr), and Antimon reinforced (Sb) was examined by Barzani et al. [5]. In the study, a TiN coated tool was used insert VBGT110302F. The cutting speed and the feed were changed in three levels (Vc=70-130-250 m/min; f=0.05-0.01-0.15 mm), while the depth of cut was held at a constant value (ap=0.5 mm). They performed 36 experiments, and they have defined the membership functions of the input and the measured output (surface roughness) parameters. Finally, they have constructed a predictive model using fuzzy approach to estimate the surface roughness.Rajasekaran et al. [6] investigated the effect of cutting parameters on the estimation of surface roughness values by using CBN cutting tool to obtain good surface quality and fuzzy modelling for turning CFRP composites.A fuzzy model to predict surface roughness and cutting power for drilling Aluminum AA1050 was used byNandi et al. [7]. Lo [8] concluded that an adaptive network based fuzzy inference system for prediction of workpiece surface roughness in end milling. Ray et al. [9] investigated the effects of cold deformation on the cutting force, surface roughness and chip characteristics in machining of ASME 12L14 free cutting steel using fuzzy set theory. A fuzzy logic model was developed by Ilkaz [10] for turning. He determined cutting parameters depending on hardness of workpiece and tool material and power of machine tool. He used Matlab Fuzzy Toolbox to prepare of fuzzy logic models.Wong et al. [11] created the general fuzzy model for the selection of cutting parameters in machining. Several fuzzy models for different cutting tools were created and compared. The results showed that the accepted model was formed with an average error of 6%. Chungchoo and Saini [12] estimated tool wear in CNC turning using the fuzzy and neural network model. For this, classification of tool wear is done with fuzzy logic, inputs are normalized, network is set to at least squared fault, and crater wear is estimated with high accuracy.

In the present work, fuzzy logic models will be developed to determine the optimum rotational speed (*n*), feed rate (*f*), depth of cut (*a*) and tool tip radius (*r*) for surface roughness. The developed model can be effectively used to predict the surface roughness in the machining of Co28Cr6Mo within the ranges of variables studied.

2. Experimental Work

2.1. *Machining conditions and roughness measurements*

In the experimental study, Co28Cr6Mo ASTM F 1537 wrought steel having hardness of 40 HRC was used as workpiece material. The dimensions of the sample are Ø50x500 mm. Turning process was carried out on a TC25-L type Sogotec CNC lathe and surface roughness values were measured on a SJ-201 Mitutoyo device. The tests were performed under dry machining conditions. The tool holder of used was MTJNR-L2525 M16, the cutting tools were TiCN coated TNMG 160404, 160408 and 160412 MT by PVD method.The average of surface roughness valuesmeasured from three different parts of the surface after each cutting operation was used. A total of 27 experiments were carried out in accordance with L_{27} (3^4) Taguchi standard orthogonal test design for four parameters and three levels used in this study. Finally, the cutting parameters were modeled with fuzzy logic as 4 inputs (*n*,*f*,*a*,*r*) and a single output (*R*a).

2.2. *Modelling of cutting parameters with Fuzzy Logic*

Limit values of the membership functions of the output parameterand input parameters are shown in Table 1. In Figure 1, membership functions and foot widths of input parameters and output parameter are shown.

Table1. Limit values for input (*n*, *f*, *a*, *r*) and output (*R*a) parameters

Rotational speed, (rpm)	Degree of speed
318-477	Slow
318-636	Regular
477-636	Fast

Feed rate, (mm/rev)	Degree of feed
0.1-0.15	Low
0.1-0.25	Middle
0.15-0.25	High

Depth of cut, (mm)	Degree of depth
0.5-0.7	Little
0.5-0.9	Middle
0.7-0.9	Much

Tool tip radius, (mm)	Degree of radius
0.4-0.8	Small
0.4-1.2	Middle
0.8-1.2	Large

*R*a, (µm)	Deg. of roughness
0.81-2.3	Excellent
0.81-3.8	Good
2.3-5.3	Average
3.8-6.75	Bad
5.3-8.207	Terrible

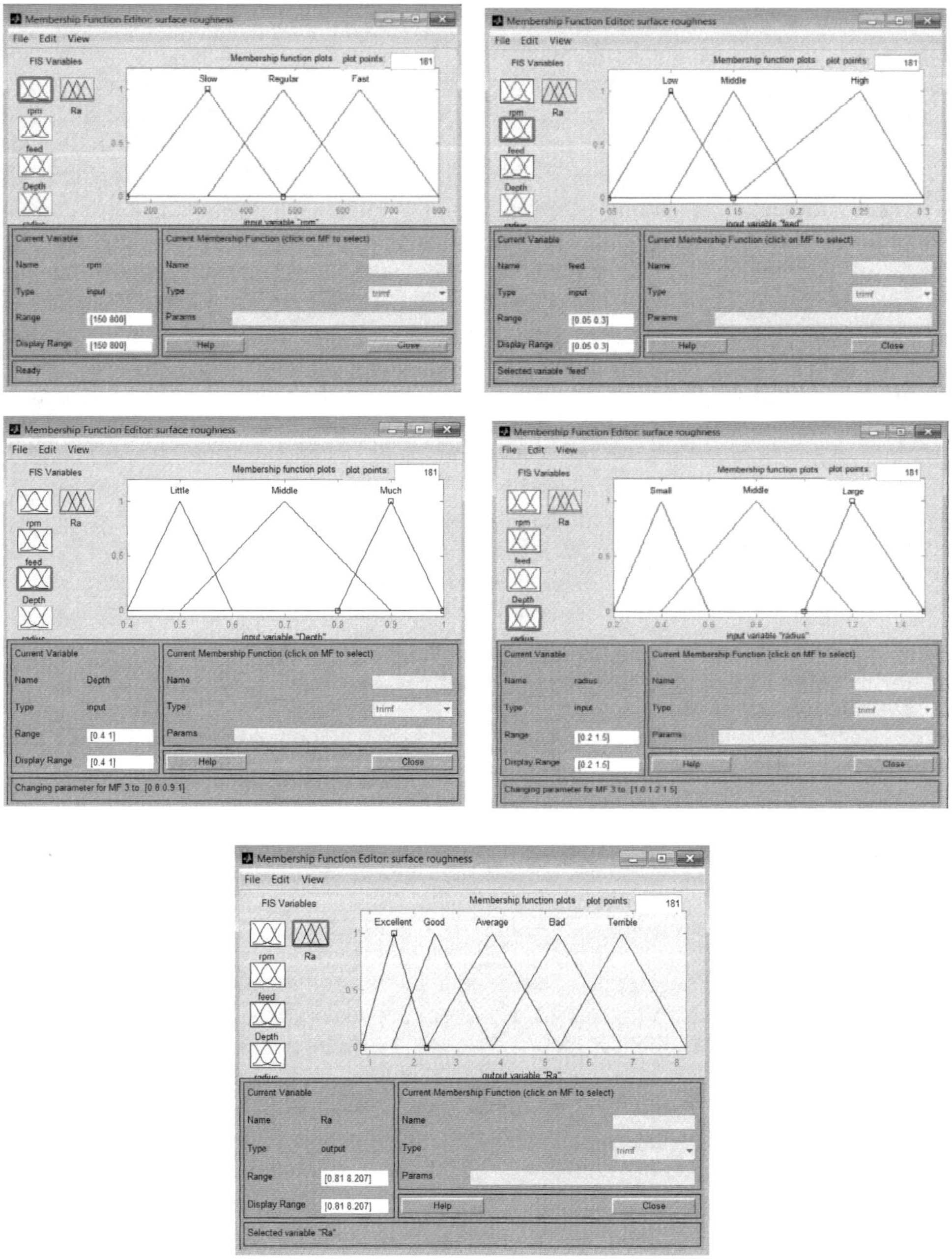

Figure 1. Membership functions and foot widths of input parameters (*n*,*f*,*a*,*r*) and output parameter(*R*a)

3. Results and Discussion

In this study, which aims to model cutting parameters with fuzzy logic in CNC turning, 222 rule bases were created to apply cutting parameters to provide the best surface quality.

3.1. Effect of cutting parameters on surface roughness

In this section, the surface roughness values of predicted are shown with 2dimensional graphics by fuzzy logic in relation to changes of the input parameters in Figure 2. For example, for n = 318, f = 0,25, a = 0,7 and r = 0,4;

IOP Conf. Series: Materials Science and Engineering **212** (2017) 012012 doi:10.1088/1757-899X/212/1/012012

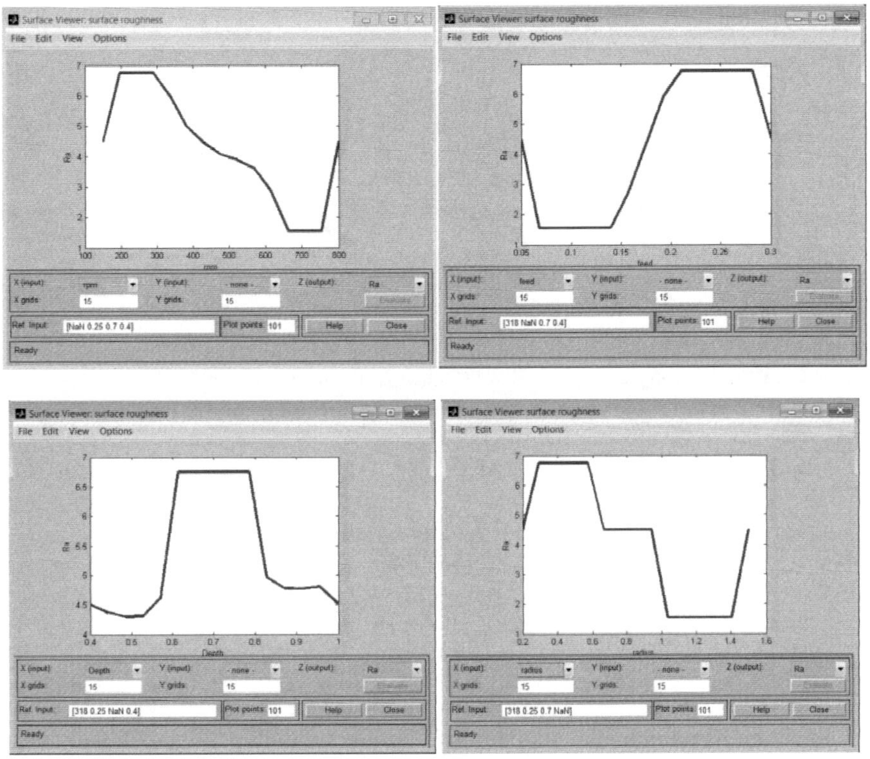

Figure 2. Effect on surface roughness of input parameter (*n*,*f*,*a*,*r*)

3.2. The comparison of experimental data and Fuzzy output

Table 2 shows the comparison of experimental data and the fuzzy output *R*a values.

Table2. Comparison of experimental and fuzzy output *R*a values.

No. of Experiment	rotational speed, *n*, (rpm)	Feed rate, *f*, (mm/rev)	Depth of cut, *a*, (mm)	tool tip radius, *r*, (mm)	Surface Roughness, *R*a (measured), (μm)	Surface Roughness, *R*a (fuzzy output), (μm)	Δ *R*a Error
1	318	0,1	0,5	0,4	1,66	2,22	0,56
2	318	0,1	0,7	0,8	0,81	1,55	0,74
3	318	0,1	0,9	1,2	1,07	1,55	0,48
4	318	0,15	0,5	0,8	1,593	2,22	0,63
5	318	0,15	0,7	1,2	1,137	1,55	0,41
6	318	0,15	0,9	0,4	2,92	2,94	0,02
7	318	0,25	0,5	1,2	2,75	2,94	0,19
8	318	0,25	0,7	0,4	7,11	6,75	0,36
9	318	0,25	0,9	0,8	4,923	4,53	0,39
10	477	0,1	0,5	0,8	1,59	2,22	0,63
11	477	0,1	0,9	1,2	0,987	2,94	1,95
12	477	0,1	0,9	0,4	1,69	3,67	1,98
13	477	0,15	0,5	1,2	0,857	2,22	1,36
14	477	0,15	0,7	0,4	4,41	4,53	0,12
15	477	0,15	0,9	0,8	2,647	3,32	0,67
16	477	0,25	0,5	0,4	8,207	4,77	3,44
17	477	0,25	0,7	0,8	3,037	3,32	0,28
18	477	0,25	0,9	1,2	1,95	2,94	0,99
19	636	0,1	0,5	1,2	1,69	2,22	0,53
20	636	0,1	0,7	0,4	4,017	3,87	0,15
21	636	0,1	0,9	0,8	1,72	2,22	0,5
22	636	0,15	0,5	0,4	3,567	3,67	0,1
23	636	0,15	0,7	0,8	1,417	1,55	0,13
24	636	0,15	0,9	1,2	2,547	2,94	0,39
25	636	0,25	0,5	0,8	3,243	2,94	0,3
26	636	0,25	0,7	1,2	2,193	2,22	0,03
27	636	0,25	0,9	0,4	4,787	3,87	0,92

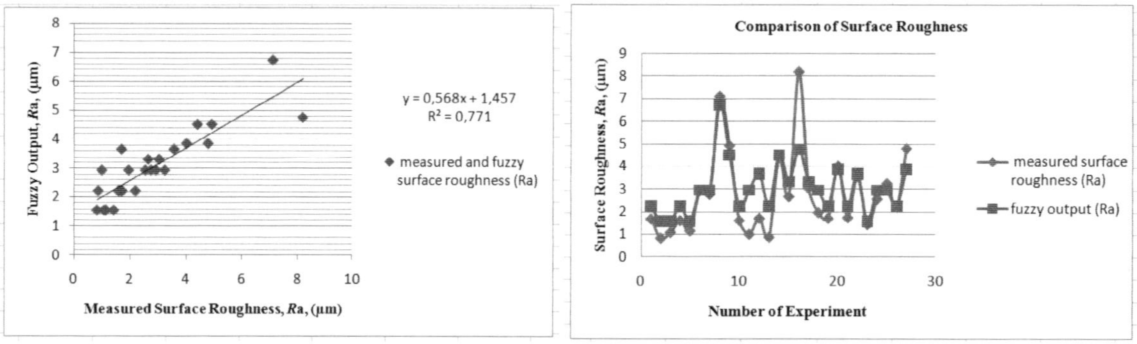

Figure 3. Comparison of theexperimental and fuzzy output results

4. Conclusions

In the present work, fuzzy logic models were developed to determine the optimum rotational speed (*n*), feed rate (*f*), depth of cut (*a*) and tool tip radius (*r*) for surface roughness.Apart from this work, the results of fuzzy logic can be compared to the results of other artificial intelligence methods such as (Neural networks, ANFIS,MANFIS, Genetic Algorithms, etc.) in the future. This work will contribute to the academic work and the manufacturing sector for predictive force model, tool wear model, tool cost, etc.

References

[1] J. Yen, R. Langari, "Fuzzy logic intelligence. Control and Information",Upper Saddle River, NJ: Prentice Hall Inc.;1999.

[2] C. Elkan, "The Paradoxical Success of Fuzzy Logic", University of California, San Diego, IEEE Expert, 1994.

[3] J. Fei, & I.S. Jawahir, "A fuzzy knowledge-based system for predicting surface roughness in finish turning", Proc. IEEE Int. Conf. Fuzzy Systems, pp. 899–906, San Diego, CA, March 8–12, 1992.

[4] A. Gok, "A new approach to minimization of the surface roughness and cutting force via fuzzy TOPSIS multi-objective grey design and RSA", Measurement, 70, pp. 100–109, 2015.

[5] M.M. Barzani, E.Zalnezhada, A.H.D. Sarhana, S. Farahanyb,S. Ramesha, "Fuzzy logic based model for predicting surface roughness of machined Al–Si–Cu–Fe die casting alloy using different additives turning", Measurement, 61, pp. 150–161, 2015.

[6] T. Rajasekaran, K. Palanikumar, B.K. Vinayagam, "Application of fuzzy logic modeling surface roughness of turning CFRP composites using CBN tool", Springer, Prod. Eng. Res. Devel., (2011), 5: 191-199.

[7] A.K. Nandi, J.P. Davim, "A study of drilling performances with minimum quantity of lubricant using fuzzy logic rules", Mechatronics, 19, (2009), 218–232.

[8] S.P. Lo, "An adaptive-network based fuzzy inference system for prediction of workpiece surface roughness in end milling", J. Mater. Process Technology, 142, (2003), 665–675.

[9] K.K. Ray, J. Das, A. Dixit, M. Chakraborty, S. Chakrabvorty, S.N. Guha, "Effect of cold deformation on the machinability of a free cutting steel",Materials and Manufacturing Processes, 2006, 21, 333–340.

[10] S. İlkaz, "Cutting condition determination for CNC turning operations using fuzzy logic",MSc. Thesis, Süleyman Demirel University, Isparta, Turkey, 2002.

[11] S.V. Wong, A.M.S. Hamouda, M.A. El Baredie, "Generalized Fuzzy Model For Metal Cutting Data Selection", Journal of Materials Processing Technology, 89-90, 1999, s. 310-317.

[12] C. Chungchoo, D. Saini, "On-Line Tool Wear Estimation in CNC Turning Operations Using Fuzzy Neural Network Model", International Journal of Machine Tools & Manufacture,42, 2002, s. 29-40.

IOP Conf. Series: Materials Science and Engineering **212** (2017) 012013 doi:10.1088/1757-899X/212/1/012013

Force and Strength Analysis of the Reel with Jaw of Torsion-bar Spring

Ke Ma, Weiqi Liu, Jiawei Wang, Le Gu

College of Mechanical& Electrical Engineering, Hohai University, Changzhou, 213022, China

make@hhu.edu.cn

Abstract. Structure characteristics and working principle of the reel with jaw of torsion-bar spring are introduced. The reel can not only eliminate the leakage risks of hydraulic jaw, but also reduce the investment cost of enterprises and improve the surface quality of the products. The static analysis of mandrel, sector plate and oblique wedge were conducted, and the main data of stress distribution and deformation were obtained, which provide a reliable theoretical basis for the design and optimization of the reel. The research results show that the external support has a great effect on the stress and deformation of the mandrel. With the increase of the weight of steel stress increases, the drum deformation increases, but the analysis of the position of maximum stress, can be obtained to drum stress and deformation is the main reason of excessive bending moment caused by heavy steel rolls. The bending moment and deformation can be reduced significantly at the end of the steel coil, which can effectively improve the service life of the drum.

1. Introduction

Cold coiler is one of the most important equipment in the production line of the cold rolled strip which can wind the long strip into a coil thereby not only increases the rolling speed, but also provides favorable conditions for subsequent production, transportation and storage. Nowadays, the steel industry is facing serious difficulties in both Europe and China, and even the world. Under this situation, the development emphasis of cold-rolling mill train should transfer from expansion the quantity to high quality, low cost, low energy consumption and greening [1-3].

Reel is the core equipment of the cold coiler, which directly affects the production of cold-rolling manufacturing line and the quality of the cold rolled strip[4-5]. In this study, a kind of reel with four wedges and configuring jaw with torsion-bar spring was used to overcome the problems of an enterprise production which include the oil leakage of the reel, non-circularity of the reel for the expansion of the path and the great wave of the tension of coiling etc. Further, Solidworks software is used for establishing the 3D model of the reel. Finally, the finite element model is established in the Solidworks Simulation module, and the static stress analysis is carried out, in addition, the strength, the deformation and the displacement of the reel are checked. On the basis of practice production, it is proved that the reel has the evident advantages which contain the simple structure, low investment cost, low pollution and high quality of the strip surface.

2. Structure characteristics and working principle of the reel

Fig 1 shows the structure of the reel which consists of mandrel, oblique wedges, sector plates, external support, expanding-collapsing cylinder (with a rotary joint) etc. Because the thickness of strip greater

IOP Conf. Series: Materials Science and Engineering **212** (2017) 012013 doi:10.1088/1757-899X/212/1/012013

than 3mm, so it is necessary to set the jaw in the sector plate. After the strip is inserted into the jaw and clamped by pre-tightening force of torsion bar, the tie rod pushes the internal wedge to the right along the axial direction through the cross-head, which is driven by the high pressure of expanding-collapsing cylinder. The internal and external oblique wedge is matched with the inclined plane, and the axial movement of the internal oblique wedge is transformed into the radial movement of the external oblique wedge. Sector plates rise slightly with the support of the external oblique wedges. The external wedge strip can fill the gaps between the sector plate, and a complete circle is formed on the surface of the inclined wedges and the sector plates for expansion of coiler. After the completion of the coiling, internal oblique wedge move toward the left along the axial direction while external oblique wedges contracted along the radial direction driven by the high pressure of expanding-collapsing cylinder. Sector plates are slowly contracted under the tension of the butterfly spring.

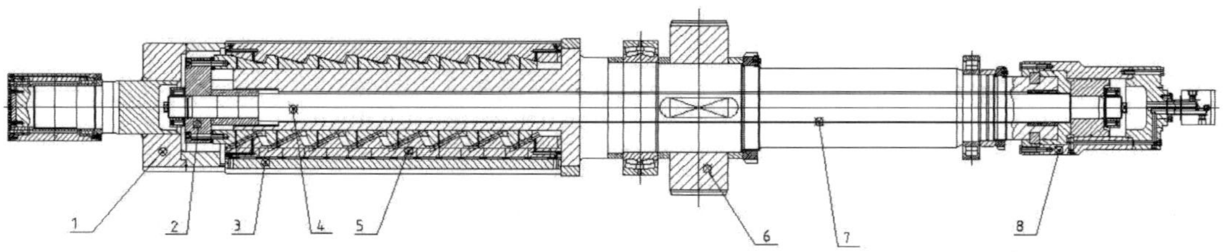

1-external support;2-crosshead ;3-oblique wedge;4-tie rod;5-jawplate;6-driving gear; 7-mandrel;8-expanding-collapsing cylinder

Figure 1. Structure of the reel

As shown in Fig 2, the torsion-bar spring type jaw is composed of tooth plate, jaw plate, hexagonal socket, torsion bar etc. It has the characteristics of simple structure, stable and reliable operation. Torsion bar spring is a torsion bar made of spring steel. When the torsion bar is manufactured, the torsion bar is made to produce a permanent torsional deformation, which has a certain pre-pressure. In order to ensure the rigidity of the structure, two torsion bars are arranged in series in the axial direction of the mandrel. In order to increase the clamping force of the jaw, surface of tooth plate is designed into a zigzag. A circular hole is processed on the oblique wedge, and hexagonal socket is arranged in the circular hole with the torsion bar. Then the torsion bar is rotated to a certain angle. When the strip is inserted in the jaw, pre-tightening force produced by torsion deformation of torsion bar can clamp the strip.

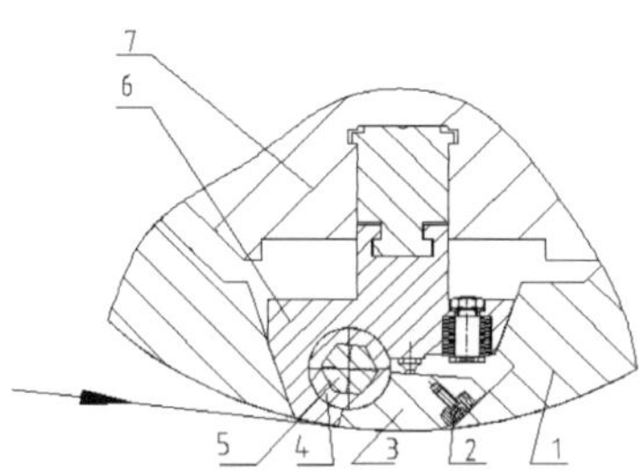

1-sector plate;2- tooth plate ;3- jaw plate; 4- hexagonal socket;5- torsion bar ;6- oblique wedge;7-mandrel

Figure 2. The structure of torsion-bar jaw

3. Finite element analysis of reel

3.1. Simulation parameter setting

As the existence of jaw may weaken the structure of mandrel, sector plates and oblique wedges, analyzing and checking the strength of reel is of great significance. Compared with the traditional mechanics of materials, the stress distribution and deformation calculated by the finite element analysis of the reel are much closer to the actual working conditions[6]. The fatigue analysis method of the main mandrel of the winding drum based on the finite element analysis is briefly introduced by Li Jisong[7]. But the mandrel in his paper is a kind of cantilever beam structure, and ignore the spool pin hole, jaw and other architecture, leads to the analysis results and the actual situation is not entirely the same. So it is very necessary to carry out finite element analysis for the reel with jaw of torsion-bar spring. The material of mandrel of the reel, sector plates and oblique wedges are all 42CrMo steel. The mechanical parameters are shown in Table 1[8]. When dividing cells, the unit grid with high quality is selected. According to the characteristics of the type and size of the reel, Solid-works stimulation can generate the corresponding curvature of the cell and automatically adjust the size and density of the cell.

Table 1. Parameters of the 42CrMo

Parameters	**Value**	**Parameters**	**Value**
Tensile trength (MPa)	≥1080	Impact energy(J)	≥63
Yield strength (MPa)	≥930	Impact toughness(J/cm2)	≥78
Elongation (%)	≥12	Hardness	≤217HB
Reduction of area(%)	≥45	Elastic Modulus(GPa)	210

3.2. Finite element analysis of mandrel

Constraint definition of mandrel: define gravity acceleration of 9.8N/m^2; define weight of steel coil of 45t; taking the external support into account, define the boss of mandrel for radial constraint; define the pressure of guide rail of P_1, P_1=3526776N which is vertical to the plane of guide rail; define the pressure of piston rod to the mandrel of S, S=466397.5N; define the bearing that near the gear for radial constraints and the bearing that near the motor for both radial and axial constraints. Finite element model of mandrel is shown in Fig 3.

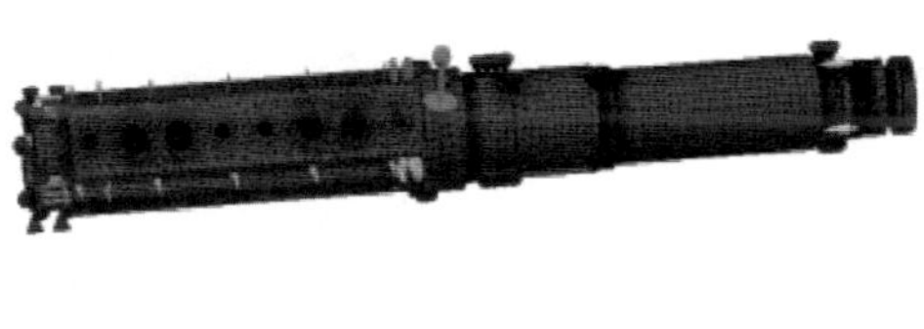

Figure 3. Finite element model of mandrel.

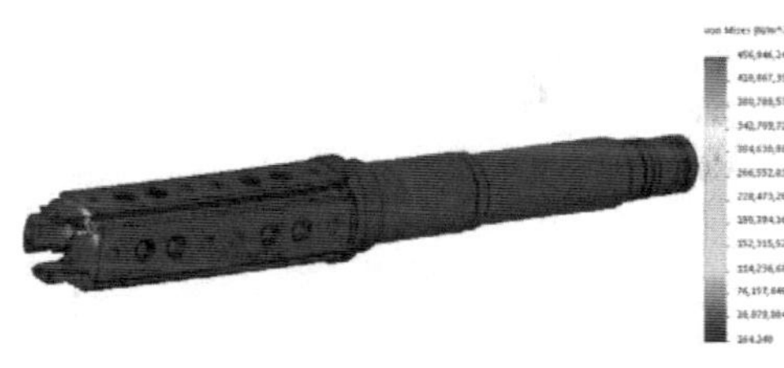

Figure 4. Equivalent stress distribution.

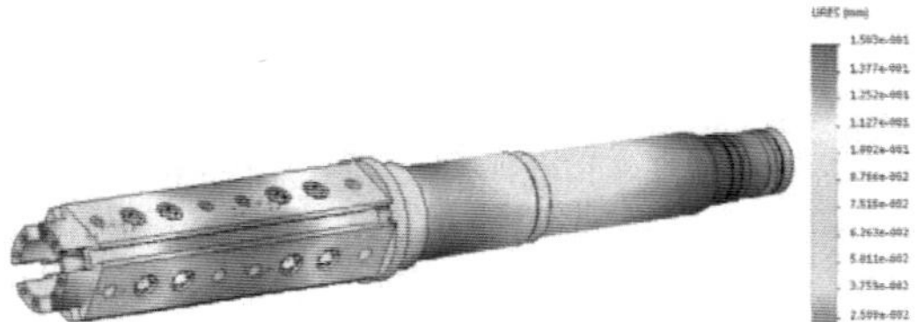

Figure 5. Equivalent displacement distribution.

Figure 6. Equivalent strain distribution.

The results of finite element model for the mandrel are shown in Fig 4, Fig 5 and Fig 6. The maximum equivalent stress occurs at the end face of the mandrel, value of which is 456.96MPa.This phenomenon is caused by pin holes near the end of the mandrel which weaken the rigidity of the mandrel. The maximum equivalent displacement occurs at the center of the guide rail, whose value is

0.1503mm and does not affect the normal operation of the reel. The reel is equivalent to the beam, the middle of which is suspended in the theory of mechanics. When the mandrel is subjected to a radial load, a larger area of stress occurs in the middle part of the beam. Therefore, the displacement of the larger area is just in line with the theory of mechanics. If the maximum equivalent stress is used to check the strength of the mandrel, the mandrel meets the requirements of the use, and the safety coefficient of the mandrel is 2.04.

3.3. Impact of the steel coil weight on the mandrel

The steel coil weight is a very important parameter and its impact on the mandrel is very necessary to be studied. It is difficult and unnecessary to establish the finite element model directly because the geometric structure of the reel is very complex which is composed of many parts. In the finite element analysis, we can grasp the main structure and ignore the secondary structure according to the main purpose of the analysis and study, and the model can be simplified without affecting the accuracy. The following assumptions were made in the model: first, the supports are rigid body without deformation; second, the reel and the steel coil are considered as a whole and the gaps and chamfers are ignored; third, the bearings were seen as a part of the support. In order to find out the influence of the external support and the weight on the stress distribution and deformation of the drum, the finite element model is established, which is shown in Fig 7 and Fig 8.

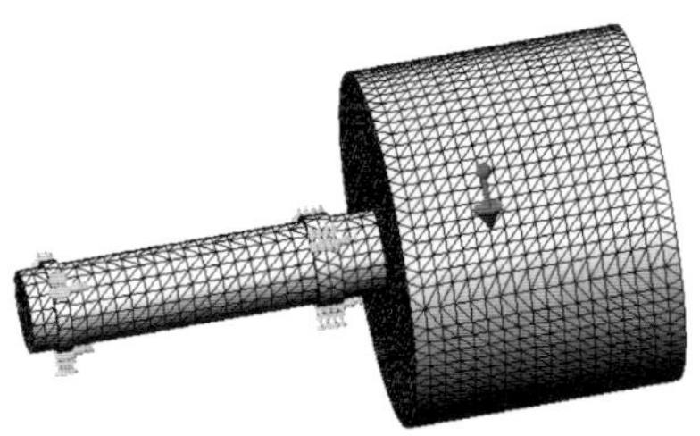

Figure 7. Finite element model of reel without external support.

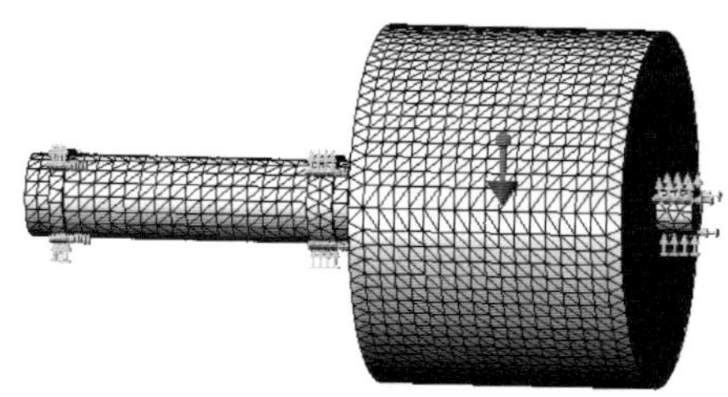

Figure 8. Finite element model of reel with external support.

The finite element analysis of the stress distribution is shown in Fig 9 and Fig 10. It can be seen from Fig 9, the maximum stress occurs at the portion of the shaft connected with the reel without considering external support. While, if the external support is to be considered, the maximum stress will occur at the portion of the shaft outside of the reel. Under different loads, the stress and deformation of the weight t=35 tons, 40 tons, 45 tons and 50 tons are calculated, and the maximum stress and deformation curves are shown in Fig 11 and Fig 12 respectively. From the results of static analysis, with the increase of the weight of steel stress increases, the drum deformation increases, but the analysis of the position of maximum stress, can be obtained to drum stress and deformation is the main reason of excessive bending moment caused by heavy steel rolls. The bending moment and deformation can be reduced significantly at the end of the steel coil, which can effectively improve the service life of the drum.

Figure 9. Reel stress distribution without external support.

Figure 10. Reel stress distribution with external support.

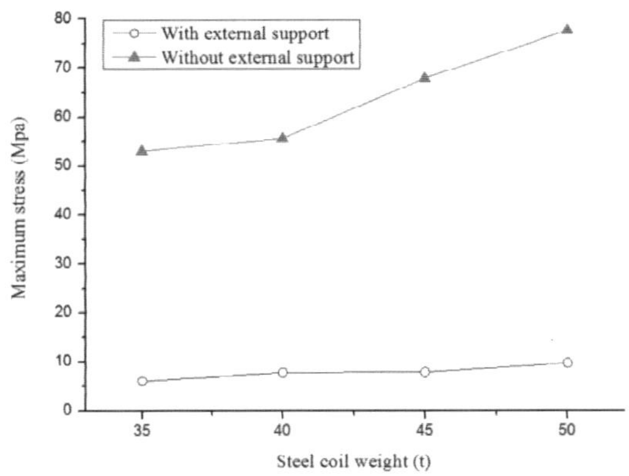

Figure 11. The maximum stress curve.

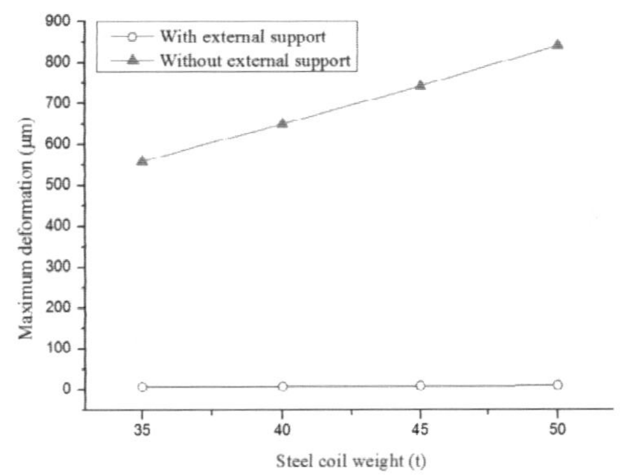

Figure 12. The maximum deformation curve.

4. Conclusions

The reel in the design has the evident advantages which contain the simple structure, low investment cost, low pollution and high quality of the strip surface. In the product design stage, the use of effective software simulation to optimize the design to improve product life is both scientific and economical methods. The finite element analysis is carried out to obtain the distribution of stress, displacement and strain, and the stress concentration position is found. According to the results of the use of enterprise site, the reel with jaw of torsion-bar spring runs normally and meets the requirements of the structure.

Acknowlegements

This research was supported by the Fundamental Research Funds for the Central Universities of China(Project No. 2013B04114) and the Natural Science Foundation of Jiangsu Province(Project No. BK20140247).

References

[1] Liu Liyan, Zhang Shutang 2010 Structure Adjustment and Development Direction of Cold Rolled Strip Production *Steel Rolling* **27** 1-6

[2] Noriki Fujita, Yukio Kimura, Koji Kobayashi etc 2016 Dynamic control of lubrication characteristics in high speed tandem cold rolling *Journal of Materials Processing Technology* **229** 407-416

[3] Yukio Kimura, Noriki Fujita, Yukihiro Matsubara etc 2015 High-speed rolling by hybrid-lubrication system in tandem cold rolling mills *Journal of Materials Processing Technology* **216** 357-368
[4] Zou Jiaxiang 2000 *The rolling and machinery* (Beijing: Metallurgical Industry Press)
[5] Tian Hongxia, WANG Feng 2008 Improvement of machining technics of coiler segment *Maintain and Rebuild*, **25** 74-75
[6] Liu C 1985 Simulation of Cold Rolling of Strip Using an Elastic-Plastic Finite Element Technique *Int. J.Mech.Sci* **27** 829-839
[7] Li Jisong 2014 Fatigue analysis of pay-off reel mandrel based on solidworks simulation *Manufacturing Automation* **10** 126-127
[8] Li YY, Zhao SD, Fan SQ, Yan GH 2013 Study on the material characteristic and process parameters of the open-die warm extrusion process of spline shaft with 42CrMo steel *J Alloys Comp* **571** 12-20

IOP Conf. Series: Materials Science and Engineering **212** (2017) 012014 doi:10.1088/1757-899X/212/1/012014

Strength of Geopolymer Cement Curing at Ambient Temperature by Non-Oven Curing Approaches: An Overview

Pitiwat Wattanachai[1], Teewara Suwan[1,2]

[1]Department of Civil Engineering, Faculty of Engineering, Chiangmai University, Thailand
[2]Center of Excellence for Natural Disaster Management (CENDIM), Chiangmai University, Thailand

teewara.s@cmu.ac.th

Abstract. At the present day, a concept of environmentally friendly construction materials has been intensively studying to reduce the amount of releasing greenhouse gases. Geopolymer is one of the cementitious binders which can be produced by utilising pozzolanic wastes (e.g. fly ash or furnace slag) and also receiving much more attention as a low-CO_2 emission material. However, to achieve excellent mechanical properties, heat curing process is needed to apply to geopolymer cement in a range of temperature around 40 to 90°C. To consume less oven-curing energy and be more convenience in practical work, the study on geopolymer curing at ambient temperature (around 20 to 25°C) is therefore widely investigated. In this paper, a core review of factors and approaches for non-oven curing geopolymer has been summarised. The performance, in term of strength, of each non-oven curing method, is also presented and analysed. The main aim of this review paper is to gather the latest study of ambient temperature curing geopolymer and to enlarge a feasibility of non-oven curing geopolymer development. Also, to extend the directions of research work, some approaches or techniques can be combined or applied to the specific properties for in-field applications and embankment stabilization by using soil-cement column.

1. Introduction

Ordinary Portland Cement (OPC) is one of the most widely used construction materials for concrete making. Its production process is mainly responsible for an enormous amount of carbon dioxide (CO_2) emission to the atmosphere [1]. Therefore, many attempts have been carried out to develop other environmentally friendly materials to achieve the concept of low carbon materials. One among those approaches is to utilise aluminosilicate (pozzolanic) wastes for the production of the OPC-less cementitious binder, which is generally known as Alkaline-activated cement or Geopolymer cement [2]. Geopolymer cement (GP) is usually produced by mixing pozzolanic materials (e.g. fly ash, bottom ash, cement kiln dust, silica fume, rice husk ash or ground granulated blast-furnace slag; GGBFS) with alkaline activators (NaOH and Na_2SiO_3). Heat curing in an oven at above ambient temperature (40 to 90°C for 6 to 48 hours) is then applied to accelerate a geopolymeric reaction and gives strength to the final products. GP sample is afterwards continually cured at ambient temperature until its further handling [3,4]. The properties of heat-cured geopolymer cement have been confirmed to be in the same order or even better than those of OPC in many aspects [5,6]. Nowadays, geopolymer can be produced on a commercial scale as precast reinforced-concrete products, but heat curing (in curing unit) is still required to achieve good mechanical properties [7-9]. To step over the limitation of using curing units and be more convenient in practical works, the development of geopolymer curing at ambient temperature is, therefore, receiving much more attention [10,11]. This paper provides an overview of the latest alternative curing methods for self-cured geopolymer cement to initiate commercial viability, on-site operation and energy saving aspect.

2. Potential factors affecting the strength of geopolymer cement cured in ambient conditions

2.1. Fineness of raw materials

An increase in surface area improves the level of both physical and chemical reactions such e.g. dissolution rate, ions transportation, forming aluminosilicate species which thereby control the initial setting time and geopolymeric gel phase [12]. The experimental data obviously confirmed that the highest strength was achieved by fine fly ash geopolymer, followed by medium-fineness fly ash and coarse fly ash respectively [13]. The similar results were also obtained by using milled fly ash or volcanic rock [14]. The full benefit of those mechanically grinding could be extendedly considered as a viable method for ambient curing with better morphology.

2.2. Mixing sequences and manufacturing procedures

Mixing order is one of the latent factors affecting properties of hardened geopolymer cement. The proper mixing order has been confirmed by many research studies to improve the mechanical properties of any alkaline-activated binder, especially for geopolymer cement [15]. The typical manufacturing process, called General mixing process, starts with the preparation of alkaline activators (e.g. NaOH and Na_2SiO_3). Raw starting materials and those solutions are incorporated and mixed at the same time [16]. A new sequence of adopting the alkaline solution, called Separate mixing process, was also proposed by mixing raw materials with sodium hydroxide solution (NaOH) before adding with sodium silicate solution (Na_2SiO_3). It was found to give better results than other sequences in mechanical properties as NaOH dissolved Si and Al from the origin and then increasingly bonded together in the presence of Na_2SiO_3 [17]. Apart from those two procedures, working with solid activators instead of alkaline solutions was also developed by just mix with called Pre-dry mixing process [18]. The main aim of this dry-mixing approach was primarily focused on the advantage in practical work with adequate strength at the ambient curing temperature (Figure 1).

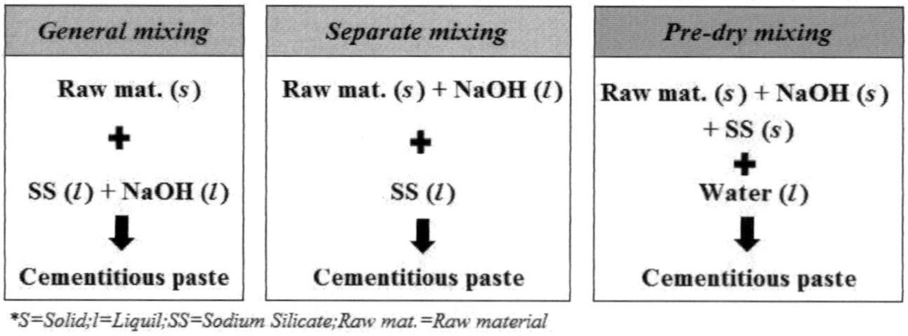

Figure 1. Different manufacturing processes of geopolymer cement

2.3. Type and concentration of alkaline activators

The most widely used alkaline solutions for geopolymer synthesis is a combination of sodium hydroxide and sodium silicate solutions due to its role in dissolve alumina-silicate minerals and initiate formation of geopolymeric gel respectively. A higher concentration gives rise to a stronger ion-pair formation, provides extra dissolution rate and complete faster polycondensation process of particle interface [19]. However, it was reported that too high concentration could lead to an increase of forming coagulated structure and obtain quick setting [20]. A combination of NaOH and Na_2SiO_3

solutions, with the concentration between 8 M to 15 M and between 30 to 50 %w/w respectively, is commonly used for the production of geopolymer cement [21]. However, the concentration level used for ambient-cured geopolymer should be significantly higher than those of typical geopolymer to obtain an extra degree of reaction [22].

2.4. The alternative heat source for curing purposes

Heat curing for geopolymer is typically carried out in electrical ovens, nevertheless, many alternative methods of geopolymer heating were observed for the best possible handling and resultant. For hot climate areas (average temperature around 36 to 42°C), the alternative heat could advocate more degree of geopolymerization to the geopolymer cement, leading to achieving higher strength. It was apparently found that the exposure condition (covered by a transparent plastic sheet and exposed to direct sunlight) was better than in shading due to an alternative heat accumulation inside the samples [23]. Self-internal heat from the massive amount of concrete pouring (huge volume) could be one of the heating sources for geopolymer concrete production. The maximum internal temperature of one cubic yard sample was around 42°C on the first day and reduced slowly to 35°C in next ten days when the ambient temperature was 25 to 30°C. It, therefore, could be applied and counted as a latent heat for geopolymer concrete using in large structure or infrastructure [24]. As aforementioned, the Pre-dry mixing process was primarily focused on the advantage in practical work at the ambient curing temperature. It was found in an internal heat measurement that dry mixing process emitted very high temperature due to the exothermic reaction when hydrating. The Pre-dry mixing method could, therefore, be developed as one of the heat based- self-cured geopolymer cement [25]. Heat from adding high potential energy compounds to geopolymer cement could be possibly counted as one of the alternative heat sources. It was reported that using CaO, $Ca(OH)_2$ or OPC as additives can shorten the setting time and raise the internal temperature inside the sample [26]. It was also found that, with Ca-based additives geopolymer, the characteristic of heat measurement was close to the general rate of OPC-heat evolution when measured with an isothermal calorimeter (mWatts/g.s or J/g).

2.5. Chemical composition and calcium content in geopolymer mixtures

It was reported that the amount of SiO_2, Al_2O_3 and Na_2O in any form of ratios apparently affects the final properties of hardened geopolymer cement. Therefore, the study of the optimum oxide ratio can be extendedly adjusted to maximize the benefit of geopolymer in ambient curing conditions. SiO_2/Al_2O_3 molar ratio is alternatively used to indicate the presence of Si and Al for geopolymeric bonding. The SiO_2/Al_2O_3 ratio of 2.0 to 3.0 is commonly prepared and resulted in excellent mechanical strength [27]. Na_2O/Al_2O_3 molar ratio and water-to-solid ratio were directly considered to affect the workability and setting time of geopolymer. An increase in sodium oxide (Na_2O) can reduce the setting time due to an efficient dissolution of aluminosilicate materials, leading to additional geopolymerization by the high level of alkalinity [28]. Na_2O/SiO_2 ratio has a role in the acceleration of the chemical shift to complex structures, rising polymeric framework. It was reported that the optimum ratio of Na_2O/SiO_2 to achieve in experiments is around 0.23 to 0.50 [29]. Calcium content in geopolymer mixtures is a key factor to control the setting time. It has been reported that partially added calcium source, such as calcium oxide (CaO), can enhance the mechanical strength of low-calcium geopolymer cement [23]. The similar results of the presence of calcium can also be found in geopolymer synthesis from e.g. high calcium fly ash, bottom ash or GGBFS as prime materials, or even the additional amount of $CaO/Ca(OH)_2$, cement kiln dust, volcanic ash containing calcium and steel slag, calcium hydroxide ($Ca(OH)_2$) or even OPC [30]. Calcium mineral could lead to the formation of C-S-H or (C,N)-A-S-H gels within a geopolymeric binder and improve the overall properties significantly and be able to develop for control the setting time of self-cured geopolymer at ambient curing temperature [10].

3. Summary of current knowledge and future research directions for the development of geopolymer cured in ambient conditions

To partially or totally replace the consumption of Portland cement, some of the alternative cementitious materials have been developed, and one among those is geopolymer cement. Most of the geopolymer properties regularly compared with OPC standard, although its formation is entirely different. The main aim of this paper is to summarize the research findings related to geopolymer synthesis on potential factors affecting its properties and characteristics when cured at the ambient temperature which is fineness of materials, manufacturing procedures, alkaline activators, alternative heat source and chemical composition in the geopolymer mixtures. The summary of compressive strength in various scenarios is presented in Table 1. It can be clearly seen that each scenario has its specific condition to achieve the highest performance. More research studies may be launched to the combination of those factors mentioned above for the feasibility of geopolymer production in ambient curing conditions. Most of the recent investigations on geopolymer were on the properties of heat curing process, which is considered as a limitation for in field applications. This study collected and provided the latest research for geopolymer cement curing without oven-heating. The major objective is to elucidate the effects of both chemical and mechanical activation on the mechanical properties of ambient temperature cured geopolymer samples. The combination of some ambient curing techniques would probably lead to skip the oven-curing process and, then be able to perform as low-to-medium strength geopolymer for other specific applications such as embankment stabilization using jet grouting soil-cement column, alternative green materials for pavement or road and infrastructure construction.

Table 1. Compressive strength of geopolymer by prime materials

Prime Materials (% wt)	Additives (% wt)	Compression MPa /Aged[a]	Alkaline Materials	Curing C°	Curing Hrs.	Sample size	Ref.
Typical OPC	-	**30.3 / 28**	-	Ambient		Cube 150 mm^3	[31]
Cement Repair	-	**46.1 / 28**	-	Ambient		Cube 40 mm^3	[32]
FA class F (90)	-	**95.0 / 28**	NaOH + Na_2SiO_3	85	20	40×40×160 mm^3	[33]
TMW (90)	$Ca(OH)_2$ (10)	**75.0 / 56**	NaOH + Na_2SiO_3	Ambient		Cube 50 mm^3	[34]
FA (80)	Water Sludge (20)	**70.6 / 90**	-	Ambient		Cube 38 mm^3	[35]
FA class F (90)	BA (10)	**70.0 / 28**	KOH + K_2SiO_3	80	24	Cube 50 mm^3	[36]
GGBFS (100)	-	**67.0 / 28**	NaOH	38	90d[a]	Cube 25 mm^3	[37]
Natural Pozzolan (100)	-	**63.0 / 28**	NaOH + Na_2SiO_3	Ambient		Cube 20 mm^3	[38]
MK (80)	Steel Slag (20)	**44.5 / 28**	NaOH + Na_2SiO_3	Ambient		Cube 40 mm^3	[32]
FA (95) class F	Silica fume (5)	**35.0 / 28**	NaOH + Na_2SiO_3	85	48	Cube 50 mm^3	[39]
Construction waste (70)	MK (20), $CaOH_2$ (10)	**26.1 / 7**	NaOH + Na_2SiO_3	80	24	Cube 25 mm^3	[23]
GBFS (75)	CKD (25)	**24.0 / 28**	NaOH	38	90d[a]	Cube 25 mm^3	[37]
Waste Paper Sludge (100)	-	**17.5 / 28**	NaOH + Na_2SiO_3	Ambient		Cube 100 mm^3	[21]

Ferronickel slag (100)	-	**15.8/28**	NaOH + Na_2SiO_3	60	24	Cube 50 mm^3	[40]
SiltyClay (75)	FA (25)	**14.0/28**	NaOH + Na_2SiO_3	75	48	Cy. dia. 50x100 mm	[41]

Note: days[a]

4. References

[1] Maholtra, V.M. (2002). Introduction: Sustainable Development and Concrete Technology. ACI Concr Int, 24(7).
[2] Al Bakria, A.M., Kamarudin, H., BinHussain, M., Nizar, I.K., Zarina, Y. and Rafiza, A.R. (2011). The effect of curing temperature on physical and chemical properties of geopolymers. Physics Procedia, 22, pp.286-291.
[3] Chindaprasirt, P., Chareerat, T. and Sirivivatnanon, V. (2007). Workability and strength of coarse high calcium fly ash geopolymer. Cement and Concrete Composites, 29(3), pp.224-229.
[4] Wongsuwan S., Paratbhumkul S., Dumneunnirun O. and Wattanachai P. (2015). A Study on the Compressive Strength and the Flexural Strength of Geopolymer Containing Ranong China Clay and Rice Husk Ash. Bachelor thesis, Chiangmai University, Thailand.
[5] Lloyd, N.A. and Rangan, B.V. (2010). Geopolymer concrete: a review of development and opportunities. 35th Conference on Our World in Concrete and Structures, August 2010, pp. 25-27.
[6] Turner, L.K. and Collins, F.G. (2013). Carbon dioxide equivalent (CO2-e) emissions: a comparison between geopolymer and OPC cement concrete. Construction and Building Materials, 43, pp.125-130.
[7] Gourley, J.T. and Johnson, G.B. (2005). Developments in geopolymer precast concrete. World Congress Geopolymer, Perth, Australia, pp. 139-143.
[8] Siddiqui K.S (2007). Strength and durability of low calcium fly-ash based geopolymer concrete, Final year Honours dissertation, The University of Western Australia, Perth 2007.
[9] Cheema, D.S., Lloyd, N.A. and Rangan, B.V. (2009). Durability of geopolymer concrete box culverts-A green alternative. The 34th Conference on Our World in Concrete and Structures, Singapore, August, Vol. 17.
[10] Suwan, T. and Fan, M. (2014). Influence of OPC replacement and manufacturing procedures on the properties of self-cured geopolymer. Construction and Building Materials, 73, pp.551-561.
[11] Singh, B., Ishwarya, G., Gupta, M. and Bhattacharyya, S.K. (2015). Geopolymer concrete: a review of some recent developments. Construction and building materials, 85, pp.78-90.
[12] Ahmari, S., Ren, X., Toufigh, V. and Zhang, L. (2012). Production of geopolymeric binder from blended waste concrete powder and fly ash. Construction and Building Materials, 35, pp.718-729.
[13] Chindaprasirt, P., Chareerat, T., Hatanaka, S. and Cao, T. (2010). High-strength geopolymer using fine high-calcium fly ash. Journal of Materials in Civil Engineering, 23(3), pp.264-270.
[14] Somna, K., Jaturapitakkul, C., Kajitvichyanukul, P. and Chindaprasirt, P. (2011). NaOH-activated ground fly ash geopolymer cured at ambient temperature. Fuel, 90(6), pp.2118-2124.
[15] Pacheco-Torgal, F., Castro-Gomes, J. and Jalali, S. (2008). Alkali-activated binders: a review. Part 2. About materials and binders manufacture. Construction and Building Materials, 22(7), pp.1315-1322.
[16] Rattanasak, U. and Chindaprasirt, P. (2009). Influence of NaOH solution on the synthesis of fly ash geopolymer. Minerals Engineering, 22, pp. 1073–1078.
[17] Hanjitsuwan, S., Hunpratub, S., Thongbai, P., Maensiri, S., Sata, V. and Chindaprasirt, P. (2014). Effects of NaOH concentrations on physical and electrical properties of high calcium fly ash geopolymer paste. Cement & Concrete Composites, 45, pp. 9-14.
[18] Suwan, T., Fan, M. and Braimah, N. (2016). Internal heat liberation and strength development of self-cured geopolymers in ambient curing conditions. Construction and Building Materials, 114, pp.297-306.
[19] Xu, H. and Van Deventer, J.S.J. (2000). The geopolymerisation of alumino-silicate minerals. International Journal of Mineral Processing, 59(3), pp.247-266.
[20] Alonso, S. and Palomo, A. (2001). Alkaline activation of metakaolin and calcium hydroxide mixtures: influence of temperature, activator concentration and solids ratio. Materials Letters, 47(1), pp.55-62.
[21] Anuar, K.A., Ridzuan, A.R.M. and Ismail, S. (2011). Strength characteristics of geopolymer concrete containing recycled concrete aggregate. International Journal of Civil & Environmental Engineering, 11(1), pp.59-62.
[22] Suwan, T. (2016). Development of self-cured geopolymer cement. Doctoral dissertation, Brunel University London.
[23] Khater, H., 2011. Effect of calcium on geopolymerization of aluminosilicate wastes. Journal of Materials in Civil Engineering, 24(1), 92-101.
[24] Vaidya, S., Diaz, E.I. and Allouche, E.N. (2011). Experimental evaluation of self-cure geopolymer concrete for mass pour applications. World of coal ash (WOCA) conference, May 2011, (pp. 9-12). [25] Suwan, T., and Fan, M. (2016). Effect of manufacturing process on the mechanisms and mechanical properties of fly ash-based geopolymer in ambient curing temperature. Materials and Manufacturing Processes (2016): 1-7.
[26] Yip, C.K., Lukey, G.C. and Van Deventer, J.S.J. (2005). The coexistence of geopolymeric gel and calcium silicate hydrate at the early stage of alkaline activation. Cement and Concrete Research, 35(9), pp.1688-1697.
[27] Burciaga - Díaz, O. and Escalante - García, J.I. (2012). Strength and Durability in Acid Media of Alkali Silicate - Activated Metakaolin Geopolymers. Journal of the American Ceramic Society, 95(7), pp.2307-2313.
[28] Ghosh, K. and Ghosh, P. (2012). Effect of Na2O/Al2O3, SiO2/Al2O3 and w/b ratio on setting time and workability of fly ash based geopolymer. International Journal of Engineering Research and Applications, 2, pp.2142-2147.
[29] De Vargas, A.S., Dal Molin, D.C., Vilela, A.C., Da Silva, F.J., Pavão, B. and Veit, H. (2011). The effects of Na2O / SiO2 molar ratio, curing temperature and age on compressive strength, morphology and microstructure of alkali-activated fly ash-based geopolymers. Cement and concrete composites, 33(6), pp.653-660.
[30] Suwan, T., Fan, M. and Braimah, N. (2016). Micro-mechanisms and compressive strength of Geopolymer-Portland cementitious system under various curing temperatures. Materials Chemistry and Physics, 180, pp.219-225.
[31] Raijiwala, D. & Patil, H., 2010. Geopolymer concrete: A green concrete. 2010 2nd International Conference on Chemical, Biological

and Environmental Engineering (ICBEE 2010), 2 to 4 Novenber 2010. Cairo, Egypt, IEEE, 202-206.
[32] Hu, S., Wang, H., Zhang, G. & Ding, Q., 2008. Bonding and abrasion resistance of geopolymeric repair material made with steel slag. Cement and Concrete Composites, 30(3), 239-244.
[33] Fernández-Jiménez, A., Garcia-Lodeiro, I. & Palomo, A., 2007. Durability of alkaliactivated fly ash cementitious materials. Journal of Materials Science, 42(9), 3055-3065.
[34] Pacheco-Torgal, F., Castro-Gomes, J. & Jalali, S., 2009. Tungsten mine waste geopolymeric binder: Preliminary hydration products investigations. Construction and Building Materials, 23(1), 200-209.
[35] Kongkaew, B., 2007. Sludge-Based Geopolymer. Master Thesis Bangkok: Kasetsart University.
[36] Hardjito, D. & Fung, S., 2010. Fly ash-based geopolymer mortar incorporating bottom ash. Modern Applied Science, 4(1), 44-52.
[37] Khater, H., 2012. Effect of cement kiln dust on geopolymer composition and its resistance to sulfate attack. International Journal of Civil and Structural Engineering, 2(3), 740.
[38] Allahverdi, A., Mehrpour, K. & Kani, E., 2008. Taftan pozzolan-based geopolymer cement. IUST International Journal of Engineering Science, 19(3), 1-5.
[39] Dutta, D., Thokchom, S., Ghosh, P. & Ghosh, S., 2010. Effect of silica fume additions on porosity of fly ash geopolymers. Journal of Engineering and Applied Sciences, 5(10), 7479.
[40] Komnitsas, K. & Zaharaki, D., 2007. Geopolymerisation: A review and prospects for the minerals industry. Minerals Engineering, 20(14), 1261-1277.
[41] Sukmak, P., Horpibulsuk, S. & Shen, S., 2013. Strength development in clay-fly ash geopolymer. Construction and Building Materials, 40, 566-575.

IOP Conf. Series: Materials Science and Engineering **212** (2017) 012015 doi:10.1088/1757-899X/212/1/012015

A Theoretical Contact Mechanics Model of Machine Joint Interfaces Based on Fractal Theory

Wenwei Liu*, Yuanhang Wang, Xiaobing Li, Chuangmian Huang, Jianfeng Yang, GuangZe Pan and Xiaojian Ding

China Electronic Product Reliability and Environmental Testing Research Institute, No.110, Dongguanzhuang Rd., Tianhe District, Guangzhou, Guangdong, China

*wenwei2009@163.com,

Abstract. To obtain more accurate contact mechanics model of joint interfaces theoretically, A theoretical contact mechanics model of joint interfaces based on fractal theory was proposed. An improved 3D WM fractal function was used to characterize the contact surface, contact load and contact area equations of asperities in elastoplastic deformation regime were established, solutions for the relationships of area-displacement and force-displacement in the elastoplastic deformation regime was done based on Hertz contact theory and fractal theory, and the present model was proven to be effective by comparing the present model to other four classical contact models and test data. Furthermore, simulations and numerical calculation results reveal nonlinear relation between the influence factors and the contact area.

1. Introduction

It is important to study and model the deformation behavior of asperities and rough surfaces using adequate parameters [1]. Hertz elastic contact theory on sphere contact was put forward for the first time in 1882, and become the basis of classical contact mechanics. Chang et al. [2-4] proposed CEB contact model based on volume conservation of an asperity, this model connected the elastic deformation and the plastic deformation, which widened the using range of Hertz contact model. Jackson and Green [5] established the finite element model on elastoplastic contact of hemispheroid, that is JG contact model. The set of cubic Hermite polynomials were used to enforce continuity between the elastic and fully plastic regimes by Brake [6], called Brake model. GW contact model was presented by Greenwood and Williamson [7] on contact behavior of joint interfaces in 1966. Majumdar and Bhushan [8] analyzed the contact behavior of the rough surface based on WM function, and this MB model opened the door to build fractal contact theory.

The main purpose of this study is to reveal the contact mechanical behavior of machine joint interfaces. Based on contact mechanics theory and fractal theory as well as considering elastic, elastoplastic and fully plastic deformations of contacting asperities, a fractal contact model of joint interfaces was proposed and studied. Moreover, the present model was compared to other four classical contact models and test data, and numerical calculation are employed to reveal nonlinear relation between the influence factors and the contact area of joint interfaces.

2. Contact model of asperities

2.1. Characterization of surface topography

Weierstrass-Mandelbrot (W-M) function [9] is widely used to characterize rough surface. A revised W-M function can be represented as

$$z(x)=L(G/L)^{D-1}(\ln\gamma)^{1/2}\sum_{n=n_l}^{n=n_h}\gamma^{(D-2)n}\left[\cos\phi_n-\cos\left(2\pi\gamma^n x/L-\phi_n\right)\right] \tag{1}$$

Where, $z(x)$ is the surface profile height, x is the lateral distance, L is the sample length, D is the fractal dimension, G is the fractal roughness parameter, γ is the scaling parameter, ϕ_n is a random phase, and n is a frequency index ranging from $n_{\min}$ which represents the lowest level corresponding to the lowest cutoff frequency to $n_{\max}$ which represents the highest level corresponding to the highest cutoff frequency.

The new size distribution function of microcontacts $n(a')$ was given as [10],

$$n(a')=D\ \ ^{(2-D)/2}a_l'^{D/2}a'^{-(D+2)/2}/2 \tag{2}$$

where, a' is the truncated area of microcontact with the radius r', a_l' is the truncated area of the largest microcontact, and   is the domain extension factor for microcontact size distribution, and the values of   is the functions of fractal dimension D.

2.2 Contact mechanics model

As is known that two rough contact surfaces can be modeled as an equivalent rough surface in contact with a rigid smooth surface. Any asperity in contact exist in one of three states, which is elastic, elastoplastic or fully plastic. The critical yield interference δ_{yc} of the asperity demarcates the transition between elastic deformation regime and elastoplastic deformation regime, and the critical plastic interference δ_{pc} of the asperity demarcates the transition between elastoplastic deformation regime and plastic deformation regime, i.e., elastic deformation of a asperity takes place when $\delta\le\delta_{yc}$, elastoplastic deformation takes place when $\delta_{yc}\le\delta\le\delta_{pc}$, and plastic deformation takes place when $\delta\ge\delta_{pc}$. The critical yield interference δ_{yc} and the critical plastic interference δ_{pc} can be written as

$$\delta_{yc}=(\pi KH/2E)^2 R,\quad \delta_{pc}=\left(\bar{p}_{pc}/\bar{p}_{yc}\right)^2\delta_{yc} \tag{3}$$

where, H is the hardness of the softer material, and K is a hardness coefficient, $\bar{p}_{yc}$ is the fully plastic indentation pressure, and $\bar{p}_{pc}$ denotes a mean pressure to initiate yield.

Therefore, the critical truncated yield microcontact area a_{yc}' and the critical truncated plastic microcontact area a_{pc}' can be, respectively, written as,

$$a_{yc}'=\left[2^{(7-2D)}\pi^{(D-1)}G^{2(D-1)}\ln\gamma(2E/\pi KH)^2\right]^{1/(D-1)} \tag{4}$$

$$a_{pc}'=\left[2^{(7-2D)}\pi^{(D-1)}G^{2(D-1)}\ln\gamma(2E/\pi KH)^2\left(\bar{p}_{yc}/\bar{p}_{pc}\right)^2\right]^{1/(D-1)} \tag{5}$$

Thus, the relationship between a_{yc}' and a_{pc}' can be derived from equation (4) and (5),

$$a_{pc}'/a_{yc}'=\left(\bar{p}_{yc}/\bar{p}_{pc}\right)^{2/(D-1)} \tag{6}$$

In the elastic deformation regime, based on the Hertz contact theory, the microcontact area a_e and the elastic contact load F_e of the contact asperities can be, respectively, expressed as

$$a_e = \pi R\delta_{yc}\left(\delta/\delta_{yc}\right)^1, \quad F_e = 2\pi KHR\delta_{yc}\left(\delta/\delta_{yc}\right)^{3/2}/3 \tag{7}$$

In the plastic deformation regime, the microcontact area a_p and the plastic contact load F_p of the contact asperities can be, respectively, expressed as

$$a_p = 2\pi R\delta, \quad F_p = Ha_p \tag{8}$$

In the elastic and plastic deformation regimes, the real microcontact area and truncated area can be connected by the following two equations,

$$a_e = a'/2, \quad a_p = a' \tag{9}$$

From equation (7), the following relationships can be deduced,

$$a_e = \pi R\delta_{yc}\left(\delta/\delta_{yc}\right)^1, \quad F_e = 2\pi KHR\delta_{yc}\left(\delta/\delta_{yc}\right)^{3/2}/3 \tag{10}$$

Similarly, the form of the above two equations can be consided in the elastoplastic deformation regime. The exponential forms are also used to expree the relationships of microcontact area a_{ep} and the microcontact load F_{ep} in the elastoplastic deformation regime,

$$a_{ep} = \pi R\delta_{yc}\left(\delta/\delta_{yc}\right)^b, \quad F_{ep} = 2\pi KHR\delta_{yc}\left(\delta/\delta_{yc}\right)^d/3 \tag{11}$$

where, b and d are constants. To derived these two constants, it is assumed that the contact area and contact load are both changing constantly in the whole stages of deformation. Therefore, based on this assumption, the boundary conditions can be established at the end of the elastic regime and the beginning of the plastic regime, i.e.,

$$a_e\big|_{\delta=\delta_{yc}} = a_{ep}\big|_{\delta=\delta_{yc}}, \quad a_{ep}\big|_{\delta_{pc}} = a_p\big|_{\delta_{pc}}, \quad F_e\big|_{\delta=\delta_{yc}} = F_{ep}\big|_{\delta=\delta_{yc}}, \quad F_{ep}\big|_{\delta=\delta_{pc}} = F_p\big|_{\delta=\delta_{pc}} \tag{12}$$

From equations (7), (8) (11) and (12), the value of b and d can be got,

$$b = 1 + \log_{\delta_{pc}/\delta_{yc}}(2), \quad d = 1 + \log_{\delta_{pc}/\delta_{yc}}(3/K) \tag{13}$$

Thus, the relationships of microcontact area a_{ep} and the microcontact load F_{ep} in the elastoplastic deformation regime can be easily obtained as,

$$a_{ep} = a'^{(b-1)(1-D)+1}\Big/\left[2a_{yc}'^{(b-1)(1-D)}\right], \quad F_{ep} = 4ER^{1/2}\delta_{yc}^{3/2}\left(a/a_{yc}\right)^{d/b}/3 \tag{14}$$

2.3. Verification of the contact model

In order to verify the validity of the present contact model, the following two methods will be used to prove it. One way is to compare present model to other classical contact model; the other way is to compare present model to the experimental data. In the present paper, the other four kinds of classical contact model including Hertz contact model, CEB contact model [2], JG contact model [5] and Brake contact model [6] are used to contrast with present model, in the meantime, present model is also compared with the test data in the literature [11]. For the convenience of comparative analysis, the critical load F_{yc}, dimensionless interference δ^*, dimensionless microcontact load F^* and dimensionless microcontact area a^* are, respectively, defined as,

$$F_{yc} = 4ER^{1/2}\delta_{yc}^{3/2}/3, \quad \delta^* = \delta/\delta_{yc}, \quad F^* = F/F_{yc}, \quad a^* = a/a_{yc} \tag{15}$$

Table 1 illustrated the experimental data of the contact sphere, the values of material properties and geometric parameters are given in Table 1.

Table 1. Material properties and geometric parameters of the contact sphere

material properties	value	material properties	value
Material name: copper		Material name: silicon carbide	
Density ρ_1 [kg/m^3]	3200	Density ρ_2 [kg/m^3]	7850
Elasticity modulus E_1 [GPa]	430	Elasticity modulus E_2 [GPa]	120
Poisson's ratio v_1	0.17	Poisson's ratio v_2	0.35
Hardness H_1 [GPa]	28	Hardness H_2 [GPa]	1.2
Yield strength σ_{y1} [GPa]		Yield strength σ_{y2} [GPa]	0.3
Radius R_1 [m]	$+\infty$	Radius R_2 [m]	0.0015

Figure 1 shown the present model compared with experimental data and other four classical contact models including Hertz model, CEB model, JG model and Brake model. Variations of the dimensionless area with the dimensionless interference is illustrated in Figure 1(a), as is demonstrated in Figure 1(a), the dimensionless area is increases along with the increment of dimensionless interference. When interference is higher, the theoretical calculating value of area in Hertz contact model is far lower than experimental data because of the elastoplastic and plastic deformation process are neglected in Hertz model, in the meantime, the theoretical calculating value of area in JG contact model is also far lower than experimental data but slightly larger than Hertz model. on the other hand, the theoretical calculating value of area in the models including present model, CEB model and Brake model agree with the experimental data. Thus, the relationship between dimensionless area and the dimensionless interference have higher precision.

Variations of the dimensionless area with the dimensionless contact load is illustrated in Figure 1(b), as is demonstrated in Figure 1(b), the dimensionless area is increases along with the increment of dimensionless load. Similarly, when contact load is higher, the theoretical calculating value of area in Hertz contact model and JG model are far lower than experimental data because of the elastoplastic and plastic deformation process are also neglected in these two models, meanwhile, the theoretical calculating value of area in CEB contact model is far larger than experimental data. These three contact models have big error, however, the theoretical calculating value of area only in the present model and Brake model agree with the experimental data. Therefore, the relationship between dimensionless area and the dimensionless load have higher precision.

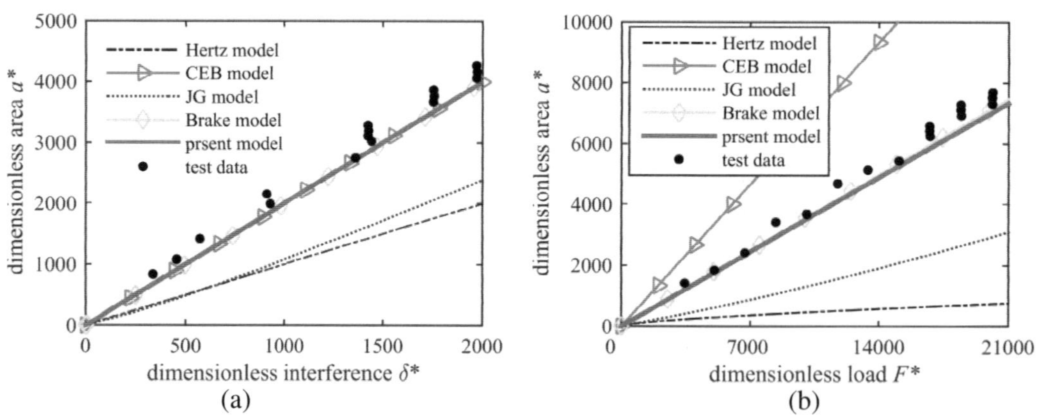

Figure 1. Present model compared with experimental data and other four classical contact models.

3. Contact model of joint interfaces

3.1. Contact load model of joint interfaces

The total load of joint interfaces F_n includes total elastic load F_{ne}, total elastoplastic load F_{nep} and total fully plastic load F_{np}. So, the total load of joint interfaces can be obtained by following calculation,

$$F_{nS} = F_{ne} + F_{nep} + F_{np} = \int_{a'_{yc}}^{a'_l} F_e n(a')\mathrm{d}a' + \int_{a'_{pc}}^{a'_{yc}} F_{ep} n(a')\mathrm{d}a' + \int_0^{a'_{pc}} F_p n(a')\mathrm{d}a' \tag{16}$$

Dimensionless total load of joint interfaces can be defined as

$$F_{nS}^* = F_n/(EA_a) \tag{17}$$

where, A_a is the nominal contact area of joint interfaces.

3.2. Contact load model of joint interfaces.

The total truncated area of joint interfaces A'_r can be got by Integral algorithm,

$$A'_r = \int_0^{a'_l} a'n(a')\mathrm{d}a' \tag{18}$$

Similarly, The total real contact area of joint interfaces A_r also includes three parts including total real elastic contact area A_{re}, total real elastoplastic contact area A_{rep} and total real fully plastic contact area A_{rp}. So, the total real contact area of joint interfaces can be obtained as,

$$A_r = A_{re} + A_{rep} + A_{rp} = \int_{a'_{yc}}^{a'_l} a_e n(a')\mathrm{d}a' + \int_{a'_{pc}}^{a'_{yc}} a_{ep} n(a')\mathrm{d}a' + \int_0^{a'_{pc}} a_p n(a')\mathrm{d}a' \tag{19}$$

Dimensionless total real contact area of joint interfaces can be also defined as

$$A_r^* = A_r/A_a \tag{20}$$

4. Simulation and calculation of Contact model

Fractal parameters and contact load are two main influencing factors to real contact area when two rough surface contact into each other. Variations of the dimensionless total real contact area A_r^* with the fractal parameters are illustrated in Figure 2, and Figure 2(a) shows the variations of the dimensionless total real contact area A_r^* with the fractal dimension *D*. As is shown in Figure 2(a), real contact area A_r^* is only a small part of nominal contact area A_a, meanwhile, there is a convex nonlinear relation between A_r^* and *D*, A_r^* increases along with the increment of *D* at first, then decreases when reaching the maximum point. The maximum point appears near the value of $D \approx 1.5$. Figure 2(b) shows the Variations of the dimensionless total real contact area with the fractal roughness parameter G^*, and A_r^* decreases along with the increment of G^*. So, the rougher of the contact surface, the larger of the real contact area, and reducing the surface roughness is helpful to increase the contact area.

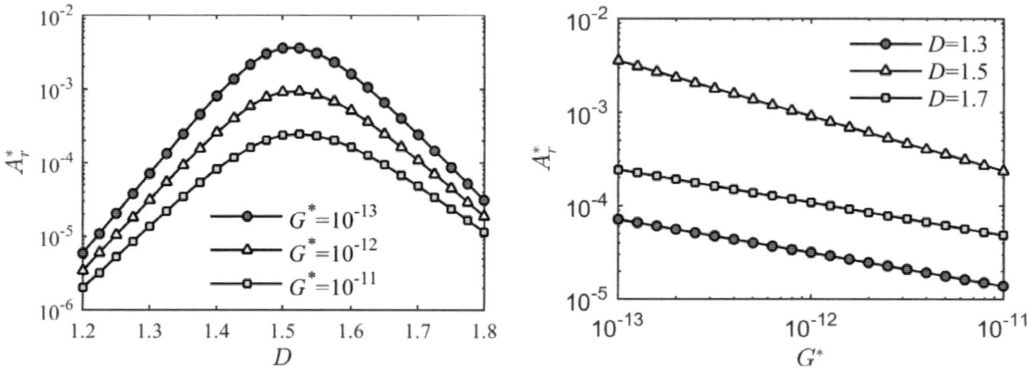

(a) (b)

Figure 2. Variations of the dimensionless total real contact area with the fractal parameters.

Variations of the dimensionless total real contact area A_r^* with the dimensionless total contact load F_n^* is demonstrated in Figure 3. As is illustrated in Figure 3(a) and (b), A_r^* increases near linearly with the increment of F_n^*, meanwhile, A_r^* increases along with the increment of G^*.

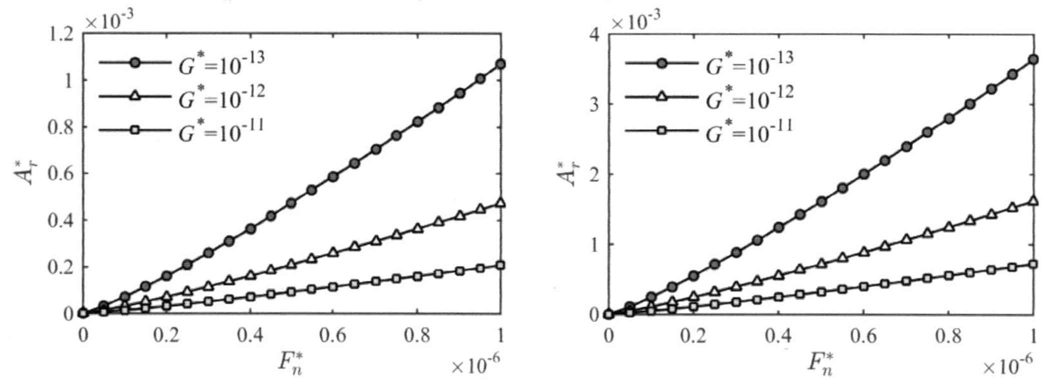

(a) D=1.3 (b) D=1.7

Figure 3. Variations of the dimensionless total real contact area with the dimensionless total contact load

5. Conclusions

Based on fractal theory and contact mechanics theory as well as considering elastic, elastoplastic and fully plastic deformations of contacting asperities, a fractal contact model of joint interfaces was proposed and studied. It was verified by comparing the present to the other four classical contact model and test data that the two basic relationships including area-displacement relationship and force-displacement relationship of asperities have higher precision. Besides, it was concluded from simulations and calculations results that dimensionless total real contact area increases along with the increment of fractal dimension at first, then decreased when reaching the maximum point, and dimensionless total real contact area decreased along with the increment of fractal roughness parameter. In the meantime, dimensionless total real contact area increased near linearly with the increment of dimensionless total contact load of machine joint interfaces.

Acknowledgment

This research was supported by Guangdong provincial research center of electronic information products reliability and environment engineering technology, Guangdong provincial key laboratory of electronic information products reliability technology, and Guangdong industrial robot reliability engineering laboratory.

IOP Conf. Series: Materials Science and Engineering **212** (2017) 012015 doi:10.1088/1757-899X/212/1/012015

References:

[1] Miao X M and Huang X D 2014 *Wear*. **309** p146~151
[2] Chang W R, Etsion I and Bogy D B 1987 *J. Tribol-T. Asme*. **109** p257~263
[3] Chang W, Etsion I and Bogy D B 1988 *J. Tribol-T. Asme*. **110** p57~63
[4] Chang W, Etsion I and Bogy D B 1988 *Journal of tribology*. **110** p50~56
[5] Jackson R L and Green I 2005 *J. Tribol-T. Asme*. **127** p343~354
[6] Brake M R 2012 *Int J. Solids Struct*. **49** p3129~3141
[7] Greenwood J A and Williamson J 1966 *Proceedings of the Royal Society of London. Series A. Mathematical and Physical Sciences*. **295** p300~319
[8] Majumdar A and Bhushan B 1991 *J. Tribol-T. Asme*. **113** p1~11
[9] Majumdar A and Bhushan B 1990 *J. Tribol-T. Asme*. **112** p205~216
[10] Wang S and Komvopoulos K 1994 *J. Tribol-T. Asme*. **116** p812~822
[11] Jamari J and Schipper D J 2006 *J. Tribol-T. Asme*. **128** p230~235

IOP Conf. Series: Materials Science and Engineering **212** (2017) 012016 doi:10.1088/1757-899X/212/1/012016

An IPSO-SVM Algorithm for Security State Prediction of Mine Production Logistics System

ZHANG Yanliang, LEI Junhui*, MA Qiuli, CHEN Xin, BI Runfang

School of Management Engineering, Zhengzhou University, Zhengzhou, 450001, China

E-mail address: junhuiemail@126.com

Abstract. A theoretical basis for the regulation of corporate security warning and resources was provided in order to reveal the laws behind the security state in mine production logistics. Considering complex mine production logistics system and the variable is difficult to acquire, a superior security status predicting model of mine production logistics system based on the improved particle swarm optimization and support vector machine (IPSO-SVM) is proposed in this paper. Firstly, through the linear adjustments of inertia weight and learning weights, the convergence speed and search accuracy are enhanced with the aim to deal with situations associated with the changeable complexity and the data acquisition difficulty. The improved particle swarm optimization (IPSO) is then introduced to resolve the problem of parameter settings in traditional support vector machines (SVM). At the same time, security status index system is built to determine the classification standards of safety status. The feasibility and effectiveness of this method is finally verified using the experimental results.

1. Introduction

Security of mine production is beneficial to the people's livelihood and social stability. Although the situation is getting better now, mine production security cannot be ignored. According to statistics, the proportion of the reported mine accidents caused by production logistics system failures reaches as high as forty percent. Therefore, the further research on mine production logistics system security has important significance in practice.

In recent years, the development of mine production logistics in academia is mainly focused on the evaluation of production efficiency[1-2], the logistics system construction [3] and the informationization. The deficiencies on mine production logistics system security are obvious. Meanwhile, great achievements have been achieved when it comes to the research of mine security status prediction, which actually have great impact on our following works. For example, X.Star Wu and Ertugrul Topuz [4] presented two Operations Research models for optimization of ventilation control device locations and sizes in two different mine ventilation systems. The using of an essentially discrete-event simulation languages has been shown to effectively simulate the behaviour of a coal transportation system in a mine which has continuous characteristics [5].YIN Dong ling [6] reexamined these factors and built an evaluation index system of mine security, with its feasibility proved by quantitative analysis. Zhu et al. [7] constructed a comprehensive classification index system of mine risk from qualitative perspective. Besides, varies of Mathematical models existed so far in mine security prediction, and the most familiar ones including "regression analysis", "Markov Model ", "Grey Model [8]", and "neural network [9]", et al. However, most models are not accurate and adaptable currently, and these studied seldom have microscopic and dynamic security state prediction of mine production logistics system.

As a promising technique of machine learning based on statistical learning theory, support vector machines (SVMs) have received widespread attention in the field of mine security for their unique advantages in solving problems related to small samples, non-linear and high dimensions. The selections of model parameters and kernel functions in the SVM have an important influence on the classification performance in reality. Considering about this, this paper puts forward an improved particle swarm and SVM optimization algorithm by using radial basis function (RBF) as the fundamental kernel function. The proposed IPSO-SVM methods maximize the mine production logistics system security state for serving the growing demand of prediction modeling.

2. The IPSO-SVM algorithm

The method proposed in the present paper introduces the IPSO algorithm to optimize the relative parameters of SVM. It combines the features of the IPSO algorithm and the advantages of SVM at the same time. Figure 1 gives the flow diagram of the IPSO-SVM algorithm.

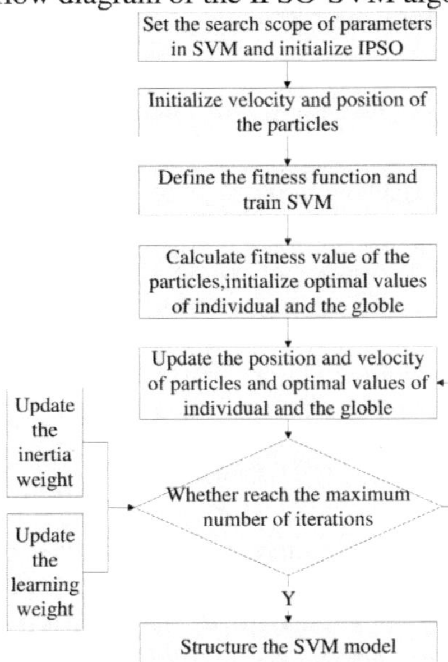

Figure 1. The flow diagram of IPSO-SVM algorithm

As can be seen, the procedure of the proposed IPSO-SVM algorithm is divided into the following seven steps:

STEP 1 Initialization of parameters. Search scopes of the parameters C and σ in SVM are set. Parameters of IPSO are initialized, including population dimension D, population size M, the maximum learning weight c_{max}, the minimum learning weight c_{min}, the maximum inertia weight W_{max}, the minimum inertia weight W_{min}, the maximum number of iterations T and the maximum velocity V_{max}.

STEP 2 Initialization of particle swarm. According to the search scope of C and σ, the model defined the initial position and speed of each particles randomly.

STEP 3 Definition of the fitness function. The average value of K models' classification accuracy is calculated by the K-CV method, which is used as a fitness function of IPSO algorithm expressed as Formula (1)

$$F(C,\sigma) = a_{K-CV} \quad (1)$$

The K-CV divides the initial data into K sets of groups. Firstly it selects a set of data as the target set, and others as the training sets. Then, each set of data will be set as a target set once one after another. The classification accuracy average of K target sets is the classification index of the SVM based on K-CV.

STEP 4 Training of the SVM. According to the value of the fitness function, the optimal value of individual species P_i and the global optimum P_g are initialized.

STEP 5 Updating the inertia weight, learning weight, positions and velocities of each particle, and optimal value of individual species and the global optimum. In details, the inertia weight is updated according to the Formula (2).

$$w = w_{\max} - (w_{\max} - w_{\min})t / T \quad (2)$$

Where $w_{\max}$ and $w_{\min}$ mean the maximum and minimum inertia weight, respectively. $t$and T are the number of iterations at present and the maximum number of iterations, respectively.

The learning weight is updated using the Formula (3).

$$c_1 = c_2 = c_{\max} - (c_{\max} - c_{\min})t / T \quad (3)$$

Where $c_{\max}$ and $c_{\min}$ are the maximum and minimum learning weight, respectively. Parameter t is the number of iterations at present and T is the maximum number of iterations.

The positions and velocities of each particle are updated on the basis of Formula (4).

$$\begin{aligned} v_i^d &= wv_i^d + c_1 r_1 (p_i^d - x_i^d) + c_2 r_2 (p_g^d - x_i^d) \\ x_i^d &= x_i^d + v_i^d \end{aligned} \quad (4)$$

Where the parameter $i = 1, 2, \cdots, m$ and $d = 1, 2, \cdots, D$, W is a nonnegative number called inertia weight, C_1 and C_2 are nonnegative constants called learning weight, r_1 and r_2 are random numbers in the range of 0 to 1. The movement velocity of every particle is set from $-v_{\max}$ to $v_{\max}$ to prevent its velocity out of control.

Finally the optimal value of individual species and the global optimum are updated by comparison.

STEP 6The machine will determine whether the iterations have met the maximum number that has been set before. If so, the iteration will stop. Otherwise, the iteration will return back to STEP 5.

STEP 7 Optimal parameter values are obtained according to the results of calculation, and the SVM model is built in the end.

3. Example and analysis

3.1. *The design of experiments*

On the basis of the index system described above, 20 sets of typical data are selected from the safety monitoring date set of operational planning underground in a coal mining enterprise in Henan.After repeated experiments and demonstration analysis, numbers from 1 to 15are chosen as the training sets, and numbers from 16 to 20 are chosen as the test sets. And theinitial experiment parameters setas:The range of penalty factor C is [0.1,100]; The range of the parameter σ is [0.1,50]; The dimension of particle D is 2; The size of particle swarm M is 100; The maximum number of iterations T is 200; The maximum learning weight $c_{\max}$ is 1.7; The minimum learning weight $c_{\min}$ is 0.9; The maximum inertia weight $w_{\max}$ is 0.9; The minimum inertia weight $w_{\min}$ is 0.4; The maximum velocity of particles $v_{\max}$ is 1.2.

3.2. *Analysis of experimental results*

In order to verify the effect on training and prediction of the IPSO-SVM model, simulation experiment is performed in the simulation environments of Matlab 2014 (the fitness functions of IPSO algorithm are as Formula (1).

In order to observe the effect of PSO algorithm after the intuitive improvement, values of the fitness function are recorded continuously during the process of iterative computations. Figures 2 and 3 are the convergence curves for the PSO and IPSO, respectively.

It is clearly seen from Figures 2 and 3 that the convergence speed and the accuracy of IPSO has significantly improved through linear adjustment of inertia weight and learning weigh. Besides, the convergence curve of IPSO is steadier compared with the counterpart of PSO, which verified the effectiveness of the IPSO to optimize parameters of the SVM.

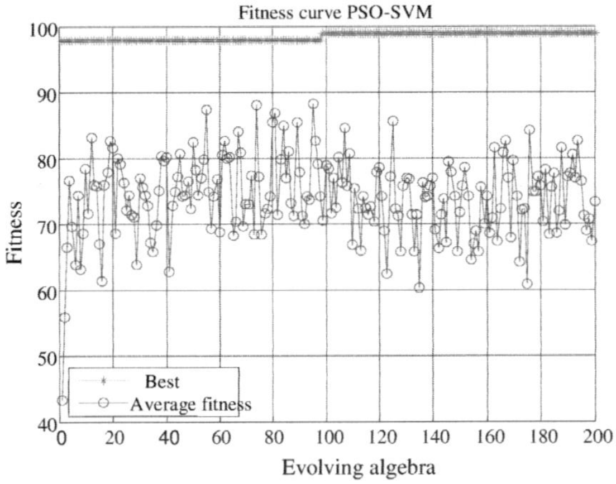

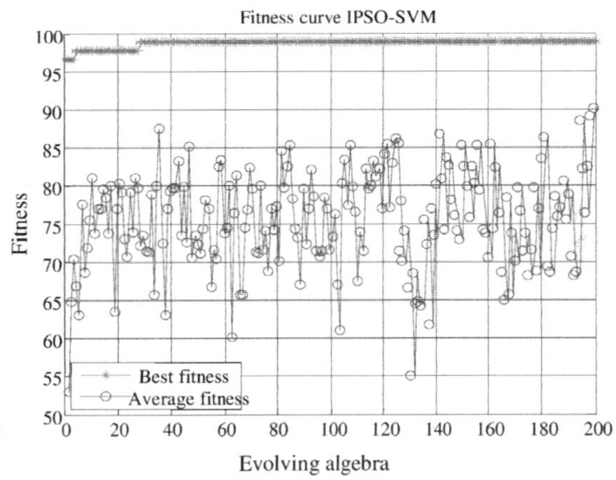

Figure 2. The convergence curve of PSO**Figure 3.**The convergence curve of IPSO

After fitting and studying the 15 sets of training samples, PSO-SVM and IPSO-SVM are used to forecast over 5 tested sample sets in experiments. The simulation and prediction results were demonstrated in Figures 4 and 5.

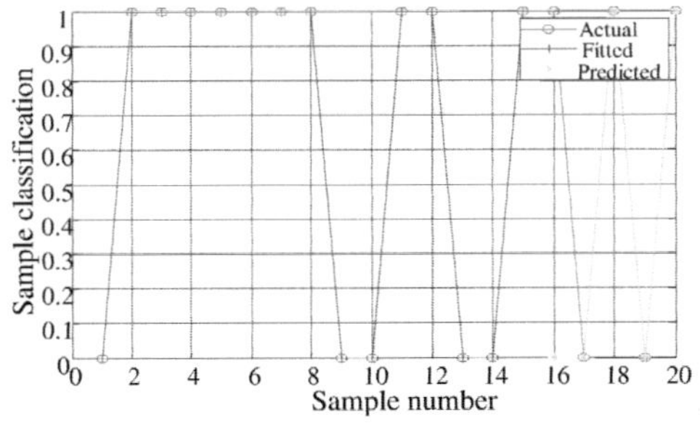

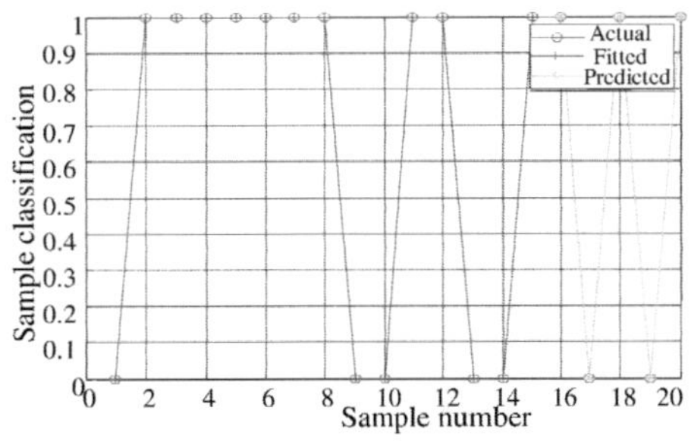

Figure 4. The classification effect of the PSO-SVM model**Figure 5.**The classification effect of the IPSO-SVM model

It is shown from Figures 4 and 5 that the PSO-VM model has one mistake in the fitting and prediction process, whilethe IPSO-SVM model in this paper made no error in the whole process. Table 1 listed all the comparisons for two models. The results demonstrates the excellent performance of IPSO-SVM model in applications, which can even characterize small sample and complicated nonlinear ones in mine production logistics system.

Table1. The comparisons of three results

Model	The number of sample	16	17	18	19	20	Accuracy	Parameters (C,σ)
PSO-SVM	Actual value	1	0	1	0	1	80%	(7.781,1.910)
	Predicted value	**(0)**	0	1	0	1		
IPSO-SVM	Actual value	1	0	1	0	1	100%	(8.181,2.350)
	Predicted value	1	0	1	0	1		

The precisions of PSO-SVM model are around 80 percent and make one error in classification, which shows the unique advantages of the SVM model in solving the nonlinear binary-class problems. In contrast, the IPSO-SVM model reaches 100 percent precision and show none error, which possesses excellent advantages in computing performance.

Above all, the experiments indicate that the IPSO-SVM is a promising model when used as security state prediction model in mine production logistics system. The model has better classification performance (none error in classification) and higher reliability (100% precision) compared with other two methods.

4. Conclusion

Combined with previous studies, The SVM theory is introduced into the prediction of security state for the characteristics of mine production logistics system. By integrating the traditional SVM model and the PSO after improvements of inertia weight and learning weight, this paper puts forward the IPSO-SVM prediction model of mine production logistics system security state. It is verified that the proposed IPSO algorithm improves dramatically the speed and accuracy of parametric optimization in the SVM. Meanwhile, the experimental results revealed that the IPSO-SVM model possesses excellent classification and prediction performance. Nevertheless, the model lacks adaptability when it comes to more complicated nonlinear problems. Therefore, the future research will focus on the introduction of nonlinear and self-adaptive parameter adjustment strategy in order to improve the search precision and speed of the particle group, and reduce the occurrence of premature phenomenon.

References

[1] Wang J F,Zhai X Q,Feng L J. Efficiency optimization of coal mine production logistics under safety hard constraint. Chinese Journal of Management Science,2014,22(7):59-66.

[2] Zhang Y L,Liu Y N,Wang J F. Research of coal mine production logistics system efficiency based on system dynamics. Industry and Mine Automation,2016,42(1):40-44.

[3] Wang J F,Hu L N,Feng L J,ea al. Dynamic coordinated development degree model for safety and efficiency of coal mine production logistics system. China Safety Science Journal,2014,24(4):15-21.

[4] X.Star Wu, Ertugrul Topuz. Analysis of Mine Ventilation Systems Using Operations Research Methods. International Transports Operation Research.1998,5(4):245-254.

[5] Edwards, J.B.Mort, N.Holland, R.J. The use of animation in the simulation of coal mine production systems[C].Third International Conference of Competitive Performance Through Advanced Technology, 2002:278-284.

[6] Yin D L,Chen Z B,Zeng J C,et al. Analysis of factors affecting coal mine operators' unsafe acts. China Safety Science Journal,2015,25(12):151-156.

[7] Zhu J F, Yang D M. Research on mine risk classification and establishment of index system. China Mining Magazine, 2011,(20):110-112.

[8] Wang L G, Pei A L. Applications of modified empennage residual error GM(1,1) model in the prediction of million tons death rate of coal mine. Science & Technology Review, 2013,(8):57-61.

[9] Li J T. Overpressure distribution prediction of mine gas explosion based on artificial neural networks. Coal Safety, 2013,(2):157-159.

Design of Grain Dryers' Control System

Li Shizhuang, Cao Shukun, Meng Wenjing
School of Mechanical Engineeri,ng, University of Jinan, Jinan 250022, China

Email: 1148660320@qq.com

Abstract: TMS320F28335 which is a TI high-performance TMS320C28x series 32-bit floating point DSP processor is used as the core of the controller, and the hardware is designed, which includes temperature collection, temperature and humidity collection, moisture detection and motor control. The development environment of the system CCS, and then for the characteristics of grain dryer control system, the control system software modular design, the use of fuzzy control method to achieve food grain motor control, and MATLAB simulation analysis, Fuzzy control is used to control the feasibility of the grain moisture.

1.Significance and background

In this essay, grain dryers' control system demands a good human-computer interface and a fast handling capability. The primary performances of grain dryers are: precision, which means moisture content of grains , which have completed drying, should try best to reach optimization value set beforehand; stability, which means in the process of controlling, the system must be stable, making sure the homogeneity of water of grain after drying; response speed, which means to every disturbance, such as the difference of grain moisture content entering the dryer, the controller should quickly make adjustment to make sure the stability of the system; adaptability, control system can normally work under a wider range of process conditions [1]. Besides, in the process of grain dryers' control, it should also reach the level of adjusting the humiture of drying medium, the temperature of grains, the temperature of exhaust gas, grain moisture content, the testing of material position and ultralimit alarming, as well as realize automatic adjustment to the temperature of dry medium and the speed of drying (the revolving speed of displace motors).

The basic aim of grain dryers' control system is to make the water of grain after drying within its safety range. The control aim of the system developed in this essay is as followed:

(1) To control the water of grain after drying by adjusting draining crop speed, with inlet grain moisture content within 14%-16% and water of grain after drying within 10-12%;

(2) To make sure grain temperature in the drying process by changing the temperature of hot wind

in the drying period;

(3) To control the process of drying work to realize the humiture of drying medium, the temperature of grains, the water temperature of modulation section, grain moisture content and the testing of material position and ultralimit alarming.

2.The general planning of system hardware

From the overall function of grain dryers' control system and the control objectives, there are several requirements about the system's controller.

(1) The performance of the control system's microprocessor must have a higher requirement. For example, in order to realize the accuracy of processing data, such as temperature, grain moisture content,etc. in real time, it must have the ability to process data quickly.

(2) The control system's microprocessor must have multi-input and multi-output interfaces;

(3) The control system's microprocessor must also have a larger and programmable storage unit and internal memory unit so as to reach the capacity of storing different types of data;

(4) The control system's microprocessor must have communication ports and multiple interrupt response methods, such as: external response, timer interrupt response,etc.

Based on above four points about the requirements to grain dryers' controller, the article primarily selects DSP as the controller of grain dryers. The general planning of the designed control system is showed in figure1.

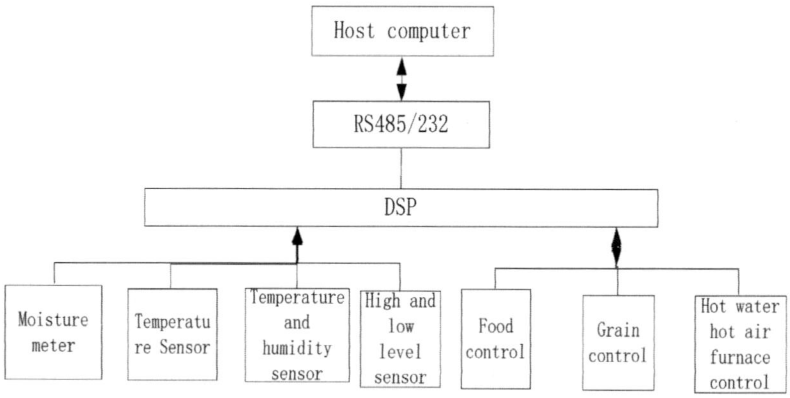

Figure 1. The overall control system block diagram

There are mainly two major parts, data collection and CMOD, of which data collection includes moisture detector and all kinds of sensors. Sensors mainly include temperature sensor, temperature and humidity sensor and material level sensor. CMOD mainly controls inlet grain elevator, drain grain motor, water and hot blast heater and other units. The distribution of all sensors in grain dryers' control system and the layout of its partial actuators are presented in figure 2.

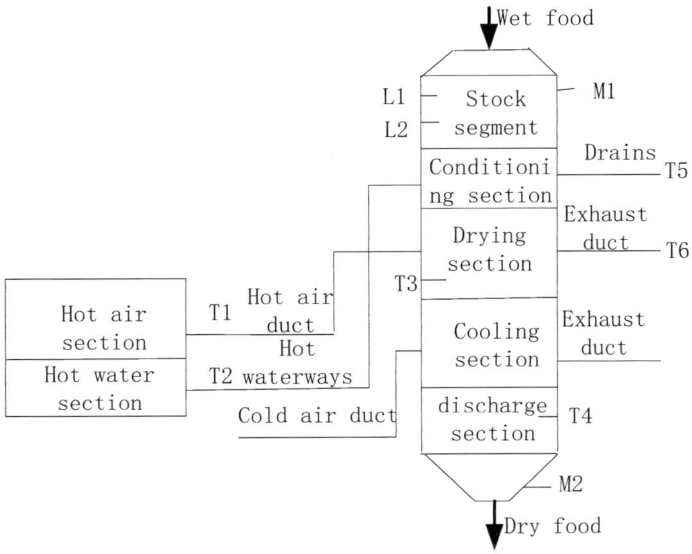

Figure 2. Grain dryer control system sensor layout

2.1 Hardware Selection

The selection of hardware is shown in the below table 1[2].

Table1.Control system hardware selection

Controller Selection	DSP(TMS320F28335)
Temperature Sensor	DS18B20
Temperature and humidity sensor	SHT11
Grain moisture detection	PT-2703
Level signal acquisition	HJ-SE Series Rotary Level Sensors

Among them, DS18B20 sensor separately tests the temperature of inlet grain, the grain temperature of drying chamber, the temperature in the cooling section and temperature of hot water.

SHT11 temperature and humidity sensor is used to test the humiture of hot wind and tail gas. The measuring instrument PT-2703 for grain moisture content is used to test the moisture content of grains at the inlet grain entrance and outlet grain entrance.

HJ-SE series damped rotation type level sensor is used to test the height of grains in grain dryers' provision section so as to fix when the elevator starts to transport grains, keep the weight of grains inside the grain dryers stable and the safety of equipment.

2.2 Electric Machine Control Module

The drying and control process of grain dryers needs a lot of electrical motors, including the feeding motor of water and hot blast heater, draught fan, water pump, drain grain motor, bucket elevator, etc. In the control process, controlling the feeding motor of water and hot blast heater and drain grain motor is very important, which can only make sure the temperature of hot water and hot wind as well as the moisture content of outlet grains during the process of drying. Hence, in this essay, the introduction focuses on the controlling of the feeding motor of water and hot blast heater and drain grain motor.

Water and hot blast heater adopts the screw conveyor to transport fuel. During the drying process, mainly rely on the screw conveyor to adjust the water and hot blast heater and control the temperature of hot wind and hot water. Adjust the motor speed of draining crop mechanism to reach the purpose of controlling the grain moisture content. Therefore, the motors of screw conveyor and draining crop mechanism adopt servo motors to drive. Compared with other controllers, one of DSP's advantages is to control the motor more easily, which is one of the most important reasons to select DSP as controller. This article adopts PWM to regulate the speed and control the transportation motor of hot blast stove[3].

Figure 3 shows the control principle of the transportation motor. DSP, as the core of control system, contains A/D conversion module inside and samples phase current to complete the corresponding current regulation. The speed regulator adopts PID control algorithm. Control the duty ratio of modulating signals according to its control output so as to control motor speed.

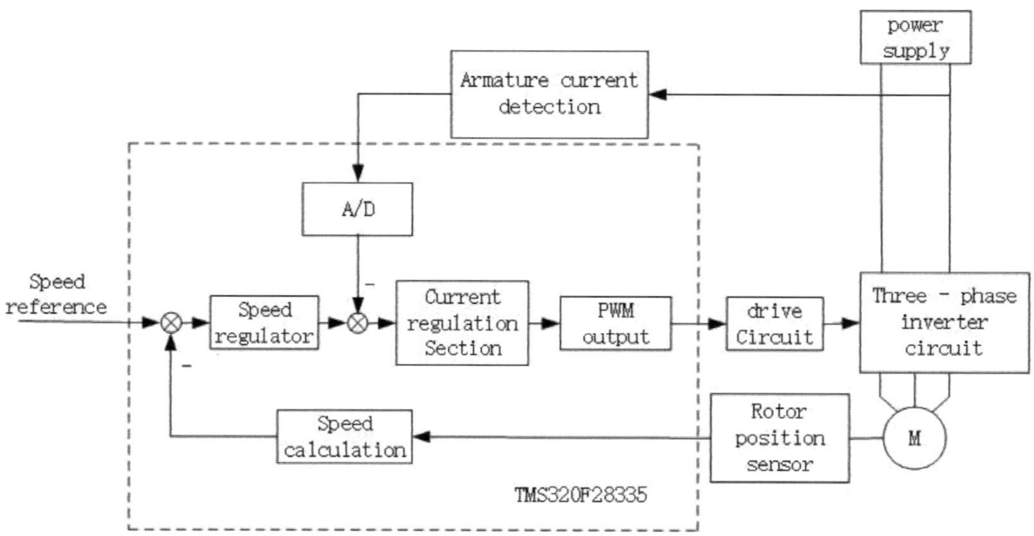

Figure 3. Motor control schematic

3.The software design of grain dryers' control system

Under the development of CCS3.3, write the control program first and then download the written program into the TMS320F28335[4].

3.1 Software's general design

The software development of grain dryers' control system adopts modular scheme, which mainly includes data collection, action executing, human-computer interface, alarm and other modules. During the control process of grain drying, mainly use the system's main program to separately adjust different subroutines to complete the functions the control system needs. The flow chart of the system's main program is shown in figure 4.

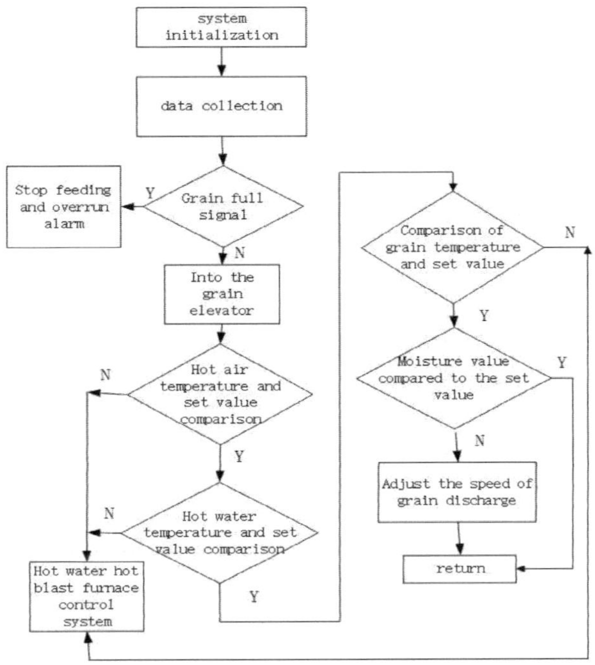

Figure 4. The main program flow chart

When the grain dryers' control system starts to work, which means DSP is being charged, firstly, invoke the system's initializer to initialize every module, such as DSP's initialization, timer and interrupt initialization, etc. When the drying process begins, collect the temperature of drying segment and cooling section inside the drying tower through DS18B20 and the temperature of inlet grain machine. SHT11 adopts humiture of hot blast air, which transfers the digital signals to DSP and compares with set value. DSP processes the adopted information to control the motor speed of hot blaster heater's inlet material machine. If the temperature of hot blast air and hot water surpasses the range, it will sound an alert. When moisture detector detects the moisture value does not match the set value, the upper computer will send out related control demands to adjust the frequency of drain grain motor and control its rotate speed.

3.2 Temperature Sample Module

DSP can control DS18B20 through a port of a line, whose specific progress is to perform reset first, and then all kinds of ROM instructions, memory reference instruction, etc. Its storage command operation table is shown as table 2, which mainly includes the command of reading serial number (33H), searching ROM commands (FOH), matching ROM (55H), skipping ROM (CCH) commands , as well as alarming and searching commands.

Table2. Stores the command table

The name of the instruction	Instruction code	Function Description
Read ROM	33H	Read the ROM serial number in the DS18B20
Search ROM	F0H	Search device encoding
Matches the ROM	55H	Finds the device based on the encoded value
Skip ROM	CCH	Skip reading the encoded value operation
Alarm search	ECH	Alarm device search

3.3 Temperature and humidity gathering module

DSP may adopt the communication mode of two-wire system to transport related data with SHT11. Before the commencement of communication, DSP firstly send start-up sequence. It means when the clock signal SCK is of high level, turn the DATA from the high level to low level and in the next time when SCK is of high level again, turn the DATA into high level. Then, invoke writing time slot to publish temperature or humidity command and delay a while to wait for the completion of measurement. During the process, SHT11 will pull the data cable into low level. When the measurement is finished, DSP will invoke reading time slots to read the measurement data of temperature and humidity and store them in the fixed units. F28335 will compare measured value with set value, producing start and stop of different equipment and alarm signals. The command of SHT11 is shown in figure 3.

Table3.SHT11 Command Table

The name of the instruction	Instruction code
Temperature Measurement	03H
Humidity measurement	05H
Read the status register	00111
Write status register	00110
Soft reset	11110

4.Fuzzy control of the drain grain motor speed

Fuzzification is to change the certain value in the discussion domain into the fuzzy sets of quantificational field. In the actual control process, the two major factors-systematic deviation and deviation variation rate play the main role. Commonly take the two amounts as the input quantities to control the system and perform adjustment. Output often chooses the amount which needs controlling. Hence, the article selects water sampling value of grain after drying and deviation value E of target

IOP Conf. Series: Materials Science and Engineering **212** (2017) 012017 doi:10.1088/1757-899X/212/1/012017

water content, the variation rate of deviation value EC, the rotate speed of drain grain motor U as linguistic variances.

In the fuzzy control of grain dryer, if set the expected value of moisture content of grain after drying as F1 and grain moisture content of drain grain entrance measured by moist content detector as F2. Then the deviation E and the change rate of deviation value EC respectively are:

$$E(t) = F1 - F2 \tag{1}$$

$$EC(t) = E(t) - E(t-1) \tag{2}$$

Select the 7-grade quantization level of three linguistic variances, which means {-6、-5、-4、-3、-2、-1、0、1、2、3、4、5、6}. The universe of discourse of deviation E is [-3, +3], the universe of the change rate of deviation value EC is [-2, +2] and the universe of U-output speed of drain grain motor is [-200, +200]. Then quantification factor Ke, Kec and scale factor Ku are respectively: $K_e = 6/3 = 2$; $K_{ec} = 6/2 = 3$; $K_u = 200/6$。

In the end, build up simulation model as is showed in figure 5.

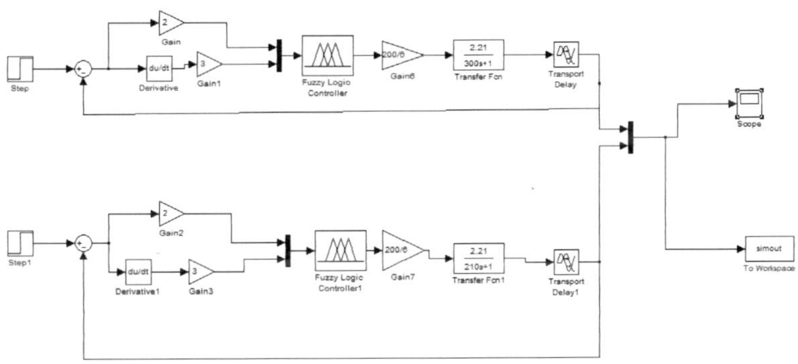

Figure 5. Fuzzy control simulation model

Simulation results are showed in figure 6. The time constant curve 1 corresponds to be 210 and the time constant curve 2 corresponds to is 300. It can be seen that the two situations can realize non-overshoot control. Besides, fuzzy control does not have clear influence on parameters variation, reduces the interference to control system by various interference factors in the process of control and applies fuzzy control onto the control of drain grain motor to make the rotate speed change steadily and fast.

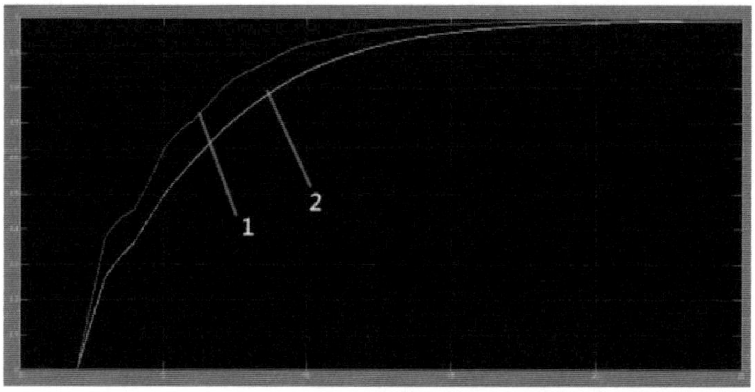

Figure 6. Simulation of step response of fuzzy control

Conclusion

Through the analysis of functional requirements of grain dryers' control system, the essay fixes the general control planning of the system. Adopt F28335 as the core controller and design the hardware, which mainly includes temperature data collection, temperature and humidity gathering, moisture detection, motor's control, etc. Ensure that adopt DS18B20 temperature sensor in the control system to measure the temperature in every part and SHT11 temperature and humidity sensor to test hot blast air. Use the PWM module of F28335 to perform frequency control on the feeding motor of water and hot blast heater and grain dryer's drain motor. In the end, it introduces to systematically develop environment CCS, and then performs modularized design for software of control system. Meanwhile, adopt the method of fuzzy control to realize the drain grain motor's control, design corresponding rules and use MATLAB to perform simulated analysis, which determines the possibility that use fuzzy control to command the outlet grain moisture.

Acknowledgement

Fund Project:

Shandong Province, agricultural machinery and equipment innovation Project "based on the efficient use of straw energy efficient integration of green energy-saving food machinery research and development and application demonstration" .

Taishan industry leading talent project special funds.

References

[1] Mei.G,Zhong.Q.W. Research on Servo Motor Control System Based onDSP[J]. Mechanical Engineering and Automation,2009,05:149-150.

[2] Xiao.Y.L,Guang.Z.C.TMS320F28335 and Its Minimum Application System Design[J]. Electronic Design Engineering,2009,01:91-92.

[3] Chalmer R E. Open-architecture cnc continues advancing. Manufacturing Engineering, 2001,126(7):48-52.

[4] Qiang.X,Ren.D.Z,Suai.Z. Design of serial communication between TMS320F28335DSP andPC[J]. microprocessor,2011,04:31-33.

[5] Shu.M.S. Research on Control System of Continuous Grain Dryer[D]. Harbin: Harbin Engineering University,2008.

IOP Conf. Series: Materials Science and Engineering **212** (2017) 012018 doi:10.1088/1757-899X/212/1/012018

Aggregate Production Planning, Case study in a Medium-sized Industry of the Rubber Production Line in Ecuador

César Rosero-Mantilla, Mayra Sánchez- Sailema, Carlos Sánchez-Rosero,Rosa Galleguillos-Pozo

Universidad Tecnica de Ambato

cesararosero@uta.edu.ec; sanchez.mayra07@gmail.com, carloshanchez@uta.edu.ec, rosagalleguillos@gmail.com

Abstract. This research aims to improve the productivity in the rubber line of a medium- sized industry by increasing the production capacities through the use of the Aggregate Production Planning model. For this purpose an analysis of the production processes of the line was made and the aggregate plan was defined evaluating two strategies: Exact Production Plan(Zero Inventory) and Constant Workforce Plan(Vary Inventory) by studying the costs of both inventory maintenance and workforce. It was also determined how the installed capacity was used with the standards of the rubber line and measures for decreasing production costs were proposed. It was proven that only70% of the plant capacity was being used so it could be possible to produce more units and to obtain a bigger market for the products of this line.+

1. Introduction

The Aggregate Production Planning strategy (APP)combines the decision making with respect of the facilities and the planning, defining the production rates formedium-term periods, strategies for the working hours (hiring, firing, subcontracting), inventories level, demand studies, inputs flow, prices and costs among other important variables for its application.[1; 2]

The main objective of the aggregate plan is to determine the optimal combination between the production index and the amount of workforce and inventory. The production index refers to the number of units produced by unit of time. The labor force level is the number of workers needed in the production and the on-hand inventory represents the unused inventory from the previousperiod.[3]

In general, the production plans of a manufacturing industry are developed through three phases: Production Plan, Master Production Schedule and the Short-term Production Planning. The short-term production plans implement the commercial policy defined in the long-term analysis based on a time horizon of one year for the assignment of resources.[4; 5]

Worldwide, the enterprises of any business area must have the workforce and technology capacity to maintain and manage efficiently the resources. The Aggregate Production Planning makes this possible and allows a higher level of competitiveness.

In Ecuador, new initiatives have been created to industrialize the processes and achieve a higher concurrence in both national and international markets with high-quality products and satisfying the demand. [7]

The enterprise analyzed on this paper is located in Ambato, Ecuador. It is a company that has many production plants such as: leather, plastic, classic canvas, relaxed-style shoe plants and also plants of thermoplastic mixtures, prefabricated products and various manufacturing industries. In the Various Industries plant many products are made, not only in the rubber line but also in the Eva line for internal and external customers. With the current demand it was found that: works were delayed, the planning was done using wrong production times, the orders couldn´t be delivered on time and finally, there was an excess demand among other problems. Therefore, it is not enough to have top-quality products.

This paper proposes a study of the production capacity to improve and/or optimize the planning and production processes in the Various Industries plant of a medium-sized enterprise in Ambato city, Ecuador.

2. Methods

Nowadays, there are many techniques to apply the aggregate production planning model. All the strategies of APP involve inventory management, production rates, levels of labor force, capacity and other controllable variables. Among the aggregate planning strategies, the most relevant are the strategies using capacity options and the ones using demand options (trying to level changes in the demand pattern occurred within the planning period). [8; 9]

APP can be established through the following steps:

Step 1:Set the demand for each period. The Master Production Schedule is given by the enterprise through the FORECAST program.

Step 2:Determine the capacity of the under time, overtime and subcontracted time of each period.

Step 3:Establish the workforce costs, hiring and firing, and also the inventory maintenance cost. The cost information was provided by the company management staff.

Step 4:Consider the company policy that is applied to the employees or to the inventory levels. In the company, only policies related to bonus payments due to increases in productivity are defined.

Step 5:Develop alternative plans and study their total costs. Two types of aggregate planning are determined by means of the mathematical models: Exact Production Plan; (Vary Workforce) and Constant Workforce (Vary Inventory and Stockout).

2.1. Exact Production Plan:

For the Exact Production Planthe variable is the number of workers, while the demand remains constant. The constraint of this plan is that the production must be equal to the demand; the equation for this type of plan is shown in equation 1:

$$Units\ produced/_{worker} = \frac{Units\ produced/_{year}}{Days\ worked/_{year}} \tag{1}$$

2.2. Constant Workforce Plan:

For the Constant Workforce Planthe variable is the production rate while the number of workers each month is a constant. The formula for the development of this plan is presented in equation 2:

$$Workers\ needed = \frac{demand/_{mont\ h}}{\left(days/_{mont\ h}\right)\times\left(units/_{worker}/_{day}\right)} \tag{2}$$

3. Results

Step1: Determine the demand for each Period. The Master Production Schedule (MPS) shows what must be produced and when the calculations and the decision making will depend on the demand. This plan included the forecast for sales and orders within the time horizon of the study, as presented in table 1 where the MPS for the next 7 months is shown.

Table1. Master Production Schedule.

MATERIAL	JUN	JUL	AGO	SEP	OCT	NOV	DIC
Front Classic carpet	410	404	414	447	466	439	366
Front Dakar carpet	2592	2558	2312	2463	2498	2488	2733
Back Dakar carpet	2329	2345	2546	2403	2360	2358	2246
Circular Flexible Rubber	129	134	123	127	129	109	125
Corrugated Flexible Rubber	1131	1178	1191	1260	1306	1017	1291
Raw Flexible Rubber	601	630	643	687	715	553	737
Mudguard	51	57	86	14	0	0	0
Mudguard layer	1150	1143	1116	1187	1202	1199	1201
Smooth Neolite	448	465	474	469	424	471	481
Diamond- tipped Neolite plus	2019	1932	1972	1992	2040	1894	1936
Total	**10860**	**10846**	**10877**	**11049**	**11140**	**10528**	**11116**

Step 2: Determine the capacity of the under time, overtime and subcontracted time for each period. The enterprise has a costing department, which is charged of studying the times for every activity and to keep them up to date in case of a change of process and/or activity. The production times in the mixing area of every product processed in each of the following lines: line of machine YT01 and line of machine GK30 were determined. The following times were also calculated: the production times in the pressing process separated by the lines of GUIX and JING DAY; production times in the sanding process in the two machines FE01 and EN06. Finally, the most commercialized group of products were determined for the study. With the established times it is possible to determine the production standards for each machine and process.

Step 3: Estimate the costs of labor force, hiring and firing, and also the costs of inventory maintenance. Table 2 shows the costs generated by each worker: direct and indirect cost; and benefits given by the company to the workers. With this information the monthly cost is obtained for the analysis.

Table2. Labor force costs.

	Monthly Income	Annual Income
Direct costs	484.73	6840.99
Indirect costs	380.72	754.28
Total	865.45	7595.27
Monthly Average	**632,94**	

*Step 4: Take into consideration the company policy applied to the workers or to the inventory levels.*As stated earlier, the enterprise has policies of bonuses related to the company productivity only for workers; no policy is applied to the inventory.

Step 5: Develop alternative plans and study their costs. For the development of aggregate plans, the enterprise supplied all the data. The number of workers in the rubber production line was of 29.

In the aggregate plan of the research the final cost is analyzed within the study time (7 months) considering the labor force costs and the inventory maintenance costs. The hiring and firing costs were not included in the analysis because internal relocation of staff can be done (from one plant to another) which doesn´t generate additional costs. The results of each aggregate plan defined in the study are presented next:

3.1. Exactproduction plan:

In this case the capacity depends on the number of workers, for this type of plan the exact demand was determined. [10]

The value of the variable is the number of staff members necessary every month to meet the demand. The constraint is that the produced units cannot exceed the demand, as it is not allowed to produce for keeping an inventory, thus inventory cost is zero.

The monthly lab or force cost is $632.94, considering direct and indirect costs. The number of units produced annually is provided by the company as well as the days worked by year. To determine the units produced by worker, the formula 1 is used and shown below:

$$Unitsproduced/_{worker} = (130729 \text{ units/year})/(7500 \text{ days worker/year}) \quad (3)$$

$$= 17.43 \text{ units produced by worker}$$

With the zero inventory plan the following results were obtained: a total cost of $123,089.08 in the seven months studied; the amount produced is exactly the demand, so the inventory maintenance cost is zero; concerning the labor force, August and October are the months that need more personnel, 29 persons and July is the month with the lowest number of workers needed, with 26 persons only. The company doesn´t have hiring neither firing costs because it implements an internal method of staff relocation, where the workers can be temporarily positioned in other plants, which means no expenses for the enterprise.

3.2. Constantworforce plan:

This plan uses the inventory produced in the low demand season to satisfy the demand of the peak demand periods and it is called level production plan because the same number of workers is used in every period. [11]

For this method, the number of workers needed during the study is determined. The demand and the working days per month are set according to the information given by the enterprise, using the formula (4):

$$Workersneeded = (76416) * (18 \text{units/worker} * 153 \text{daysavailable}) = 27{,}74 \text{ Number of workers} \quad (4)$$

By having a constant number of workers, an inventory is generated. To determine the cost of keeping an inventory, it was considered: the use of machinery and electricity, the salaries of the workers and also a reference value of the maintenance of the electric vehicles, which was $0.28 by month, almost insignificant for calculations.

With this method the results are: the cost is $125,897.74in the seven months of the study; the constant number of workers is 28 persons, therefore, in the first month one person would have to be relocated because there were 29 workersat the beginning of the study. At the end of the seven months an inventory of 696 units is generated. Finally, to determine the use of the installed capacity, the values of the demand in the studied period versus the possible production rate are considered according to the standard time, as shown in table 3.

Table3. Use of the installed capacity.

Months	JUN	JUL	AGO	SEP	OCT	NOV	DIC	Total
TotalDemand	10860	10846	10877	11149	11140	10528	11116	76416
Production Machines GUIX 01-02 and JD01-02	15120	15120	15120	15120	15120	15120	15120	105840
% Use	**71.83%**	**71.73%**	**71.94%**	**73.08%**	**73.68%**	**69.63%**	**73.52%**	**72.20%**

It is shown that 72% of the installed capacity is being used in the company. If the decision to sell more products than the ones mentioned in this analysis is considered, the possibility of having more than one work shift could be an alternative.

4. Discussion

According to the previous results, it was found that: with the Exact production plan the total cost is $123,089.07 which only includes the work force cost as no inventory is generated; and that the maximum variation value of the workforce was of 3 persons in the months of production peak.

On the other hand, with the Constant workforce plan the cost is $125,897.74, which includes labor force costs and inventory maintenance costs. 29 workers were available initially and the constant labor force is of 28 persons, that´s why at the beginning one person had to be relocated. Regarding the inventory maintenance costs, this alternative takes into consideration the Various Industries Logistics area, the wages of the personnel, the electricity and the machinery to estimate an approximate value of the cost to maintain one product in stock.

5. Conclusions

It is concluded that a study was made on the production capacity for the plant of Various Industries in the rubber line, analyzing the production processes of the line and determining the critical processes in the rubber production.

The impact of the Master Production Schedule on the production capacities was analyzed, and this made possible to propose a solution of production volume according to the capacities obtained.

The plant of Various Industries in a medium-sized enterprise in Ambato, Ecuador, has an installed capacity that is largefor the present demand of rubber products. Approximately 70% of the capacity is being used to make the current products and satisfy the current demand.

References

[1] Buxey, G. Aggregate planning for seasonal demand; reconciling theory with practice. International Journal of Operations & Production Management (2005) Vol. 25 Iss: 11, pp.1083 - 1100. http://dx.doi.org/10.1108/01443570510626907

[2] Boiteux, O., Subias, A., García, A.; La Planificación Agregada como Instrumento Integrador de las Áreas Funcionales de la Empresa; 4th International Conference on Industrial Engineering and Industrial Management XIV Congreso de Ingeniería de Organización Donostia- San Sebastián , September 8th -10th 2010, págs. 1299-1308;

[3] Chase, R., Jacobs, R., Aquilano, N., 2009, Administración de la producción y operaciones y cadena de suministro, 12a. ed., México: Editorial McGraw Hill.

[4] Das, B.P., Rickard, J.G., Shah, N., Maccietto, S., 2000. An investigation on integration of aggregate production planning, master production scheduling and short-term production scheduling of batch process operations through a common data model. Computers and Chemical Engineering, 24: 1625-1631.

[5] Chase, R., Aquilano, N. Jacobs, R.; Production and Operations Management; 8th ed, Boston, Mass.; London: Irwin/McGraw-Hill.

[6] William J. Stevenson, Operation Management, 8th edition, McGraw-Hill Book Company.

[7] Reyes, J., Molina, C; Plan Agregado de Producción Mediante el Uso de un Algoritmo de Programación Lineal: Un caso de Estudio para la Pequeña Industria; Revista EPN, Vol. 34, N°. 1, Octubre 2014.

[8] Naranjo, E.n, 2009, Análisis de la evolución histórico–cultural del proyecto de desarrollo comunitario en la Parroquia Salinas de la provincia deBolivar, Revista Politécnica, Vol. 30(1): 118–124.

[9] Heizer, J. A. y Render, B.,2004, Principios de administracion de Operaciones. Pearson Madrid.

[10] Del Solar, R., Chacón, I., Ponce, M., 2008; Plan agregado de producción en barracas madereras. Estudio de caso para una pequeña industria; Maderas. Ciencia y tecnología 10(2): 77-92.

[11] Vollmann, T., 2008, Planeación y Control de la Producción, Administración de la Cadena de Suministro, Bogotá: Mc Graw Hill.

Continuous Flow in Labour-Intensive Manufacturing Process

Eng. Jhonny Pacheco[1], MSc. Eduardo Carbajal[1],Dr.-Ing. Cesar Stoll[1]

[1]Pontifical Catholic University of Peru, Industrial Engineering, Lima-Peru

jpachecoe@pucp.pe ;ecarbajal@pucp.pe; cstoll@pucp.edu.pe

Abstract. A continuous-flow manufacturing represents the peak of standard production, and usually it means high production in a strict line production.Furthermore, low-tech industry demands high labour-intensive, in this context the efficient of the line production is tied at thejob shop organization.Labour-intensive manufacturing processes are a common characteristic for developing countries. This research aims to propose a methodology for production planning in order to fulfilment a variable monthly production quota. The main idea is to use a clock as orchestra director in order to synchronize the rate time (takt time) of customer demand with the manufacturing time. In this way, the study is able to propose a stark reduction of stock in process, over-processing, and unnecessary variability.

1. Introduction

An international curtain maker in Lima-Perú faces up the challenge to improve the delivery time to customer.As many business, the curtain market in Peru is very competitive, i.e. low prices, high quality, short delivery time. The philosophy and techniques of lean manufacturing is a need in this business environment, especially if companiescare not onlyto keep their customer but also increase their sales opportunities.

The proposed methodology is applied in the assembly line of curtains. Nevertheless the potential application is for all relatedsectors with manual assembly, labour-intensive and not much machines. Usually, industry of textile, footwear, leather, etc. has these characteristic in developing countries.In this context, the production planning task is the most benefited, because it allows working with customers under the perspective of Just in Time concepts. No more unnecessary overtime, defective products, high inventory, etc.

Theidea of this project started in Brazilian filial of the curtain maker. Here, we will present the result of adequate procedure and methods to the Peruvian reality. The result is a flexible methodology that is easy to apply to other environments.

2. Literature review

According to Womack [1], lean manufacturing is based in five principles: value, mapping, flow, pull, and perfection. That means, value to understand client perspective, mapping to identify what we really do, flow to create an ideal future state map, pull in order to match with the customer demand, and finally perfection to search the continuous improvement.

Lean operations show common characteristics, like elimination of waste, produce what and when is required, no mistakes, tighten links in the value chain, and robust processes [2]. Many of the lean techniques are simple and common sense. Nevertheless, most of the companies that tries to implement lean techniques fail. A large survey conducted by Industry Week in 2007 found that only 2% of companies that have a lean program achieved their anticipated results [3]. In July, 2010 Robert Miller, Executive Director of the Shingo Prize was interviewed on radiolean.com, he affirmed that a very large percentage of companies, that had received the Shingo Prize, had not sustained their progress after winning the award [4]. How to face this reality?

According to Spear [5], we need to decoding the DNA of the Toyota production system. In short, managers do not tell employees how to do their work, rather they use a teaching and learning approach that allows their workers to discover the rules as a consequence of solving problems. For example while a persons is doing his job, ask a series of questions: how do you do this work?, How do you know your are doing this work correctly?, How do you know that the outcome is free or defects?, What do you do if you have a problem?. In summary, the workers must be total involved by the kaizen philosophy, continuously improvement.

Our study, was based on the kaizen philosophy, and it was a challenge due to the demand of high labor-intensive, in a developing country. The one-piece flow principle was wide applied in the next case study, in order to identifying and eliminating the seven type of waste: overproduction, transporting, motion, waiting, processing, inventory, and defects. The one-piece flow principleproposes that single parts are transferred between different operations instead a complete lots [6]. That means, products that move continuously through the processing steps with minimal waiting time in between, and the shortest distance traveled, will be produced with the highest efficiency [7].

3. Methodology

This researchuses an assembly curtain line as case study. The assembly line demands labor-intensive and some no-complex machines. Five steps are applied as methodology for synchronize the customer demand with the manufacturing time:

*First Step:*Estimation of the forecast demand. For our case study, we take the historical demand of the last 57 month for a product. The forecasting demand for the year 2017 is presented in the Tabla 1.

Table 1. Forecasting Demand.

Month, year	X(t)	Y(t)	Seasonality Rate	Forecast
Jan, 2017	60 month	3157 unit	0.96	3027 unit
Feb, 2017	61 month	3198 unit	0.84	2681 unit
Mar, 2017	62 month	3239 unit	1.10	3569 unit
Apr, 2017	63 month	3279 unit	1.02	3337 unit
May, 2017	64 month	3320 unit	0.97	3210 unit
Jun, 2017	65 month	3361 unit	0.98	3292 unit
Jul, 2017	66 month	3401 unit	0.98	3336 unit
Aug, 2017	67 month	3442 unit	1.03	3537 unit
Sep, 2017	68 month	3483 unit	1.12	3912 unit
Oct, 2017	69 month	3524 unit	0.97	3430 unit
Nov, 2017	70 month	3564 unit	0.84	2998 unit
Dec, 2017	71 month	3605 unit	1.19	4293 unit

Second Step: It is necessary to fix the number of working days per month, and estimate the real work time per day. We need to identify all kind of activities that are realized during the working day, and differentiate activities that add value to the product from others. The Table 2 takes, as example, the forecasting demand for January 2017 (3027 units). The table shows the kind of daily activities and used time. For the case, it is considered that January 2017 has 22 production days.

Table 2. Estimated Working Time per Day.

Activities	Time	Percentage
Work Time	7.87 hours	82.9%
Meetings	0.05 hours	0.5%
Clean	0.08 hours	0.9%
Breaks	0.50 hours	5.3%
Lunch	1.00 hours	10.5%
Total	9.50 hours	100.0%

Third Step: Calculation of the takt time for the selected month production. For our case, we have an estimated demand for January 2017 of 3027 units (see Figure 1), 22 work days in this month, and 7.87 hours/day as real work time (see Table 2).

$$Takt\ Time\ = \frac{Work\ Time}{Demand} = \frac{22\ days * 7.87\frac{hours}{days} * 3600\frac{seconds}{hours}}{3027\ units} = 206\frac{seconds}{unit}$$

The calculated takt time indicates that line must produce a finished unit every 206s, in order to fulfilment the demand of January 2017.

Fourth Step: Calculation of the cycle time. In this step, we identified all processes needed in order to finish the product. A time and motion study must be carried out in order to improve the efficientand better opportunities to control the processes variability. A resume of this step for the case study is presented in the Table 3.It is identified 39 processes and 820.2 seconds to finish one product.

Table 3. Resume of the Time and Motion Study – Cycle Time.

Process	P1	P2	P3	P4	P5	P6	…	P34	P35	P36	P37	P38	P39	Cycle Time
Time (s)	10.4	23.8	14.2	17.4	24.8	14.4	…	21.2	21.8	25.3	19.4	22.2	19.2	820.2

*FifthStep:*Defining the number of workstations. We need to synchronize the cycle time (820 s) with the required manufacturing time. Thus, it is possible to fulfillment the monthly demand. For the case, it means that the line production must have a takt time of 206 seconds (step 3).

$$Number\ of\ Stations\ Work\ required\ = \frac{Cycle\ Time}{Takt\ Time} = 3.98 \cong 4\ Work\ Stations$$

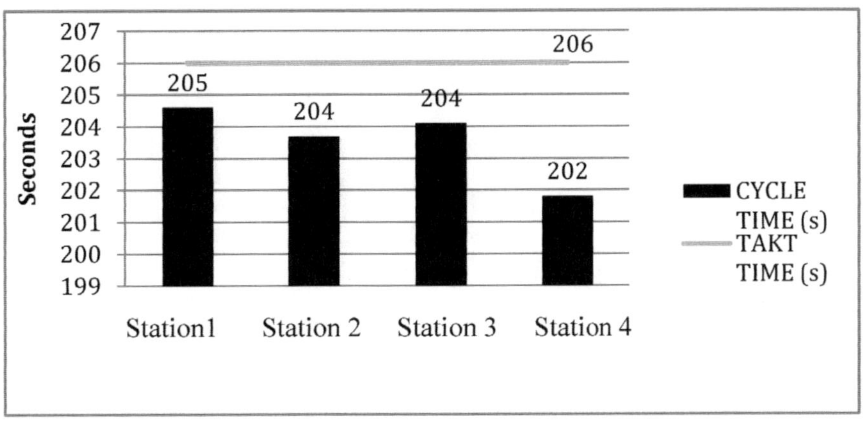

Figure 1. Synchronization Cycle Time with the Takt Time.

We need to group the processes (table 3) in workstations. Some processes could keep together due to production constrains. The Figure 1 presents the four stations required for the case study. The workstation 1 includes from process 1 (P1) to process 11 (P11), the cycle time of this workstation is 205.6s. According to the figure, the standard deviation is 0.75s for the cycle time of each workstation.

After the five steps, we propose the use of a clock as orchestra director for the line production. The clock is an effective tool to keep the rate of the production, i.e. synchronizing the cycle time with the takttime.

The figure 2 shows an example of the kind of clock required. Each time, that the clock completes its cycle, one finished product must be exit from the productionline. The cycle of the clock must be adjusted according the takt time (step 3). Each workstation must follow the rate of clock in order to fulfilment the production plan of the month. The figure 2 not only shows the cycle but also the quantity of products that must be produces in the day, and the remained estimated time needed to reach the goal production. This kind of information helps the workers to keep the rate production.

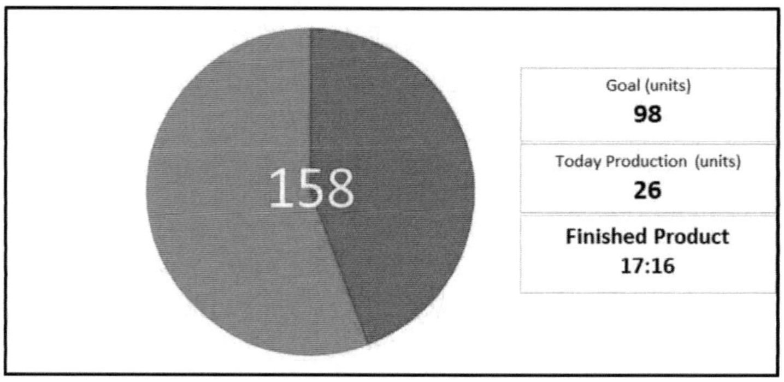

Figure 2. Clock as Orchestra Director.

4. Results and discussion

The figure 3 shows results of the applied methodology during one day work in January 2017. The average takt time was 205.5 seconds with a standard deviation of 1.7 seconds.

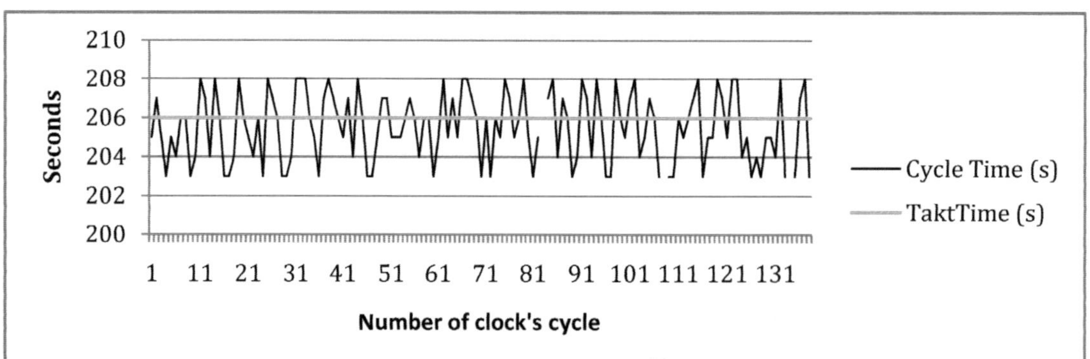

Scheduled	138 unit
Real	135 unit
Workstation 0	3 unit
% Fulfilment	100%

Figure 3. Real Cylce Time vs Takt Time.

We can also notice that the goal production was achieved but it was necessary a "Workstation 0". During the day production, three units faced some kind of problems and must been finished in a parallel workstation in order to not stop the line production. The takt time line, in figure 3, has three disruptions.

IOP Conf. Series: Materials Science and Engineering **212** (2017) 012019 doi:10.1088/1757-899X/212/1/012019

The Table 4 presents the summary results, and they are favourable spite of the forecast error (25 units). The demand was attended (3002 units), no client complains due to quality problems or incomplete orders. Before the firm applied this methodology, they could not attend the demand and needs to pay overtime.

Table 4. Summary Results (January 2017).

Forecast Demand	3027 unit
Real Demand	3002 unit
Real Production	3036 unit
Stock	34 unit

5. Conclusions

A high efficient synchronization between rate time of customer demand and the manufacturing time is possible, if we have a good forecast demand and a time and motion study. These studies allow estimating a good takt time and an appropriate number of workstations.

The proposed clock could be controlled in a web application, and it must be localized in order to facilitate the read along the production line. A called "Workstation 0" allows facing up eventual problems in the line production. Due to our case study is based on labour-intensive, manual assembly and not much machines, the proposed workstation is not expensive. At the beginning, the use of a "Workstation 0" is high, but as the workers come to be on better terms, the need of a "Workstation 0" decreases drastically.

References

[1] Womack, J., Jones, D., Lean Thinking: Banksh. Waste and Create a Wealth in your Corporation. New York: Simonand Schuster (1996)

[2] Miller, J., Wroblewski, M., Villafuerte, J., Creating a Kaizen Culture - align the organization, achieve breakthrough resutls, and sustain the gains. McGraw-Hill (2013)

[3] Industry Week, Everybody's Jummping on the Lean Bandwagon, but Many are Being Taken for a Ride (2008)

[4] Liker, H.., Rother, M., Why Lean Programs Fail. Lean Enterprise Institute (2011)

[5] Spear, S., Bowen, K., Decoding the DNA of the Toyota Production System. Harward Business Review, pp. 95-106 (1999)

[6] Nyhuis, P., Vogel, M., Adaptation of logistic operating curves to one-piece flow processes. International Journal of PROductivity and Performance Management, 55(3/4), 298-299 (2006)

[7] Liker, J., Meier, D., The Toyota Way Fieldbook: A practical guide for implementing Toyota's 4Ps. Pp. 80-110. McGraw-Hill (2006)

Finite Progressive Planning for the Assembly Process in Footwear

John Reyes[1], Darwin Aldás[1], Edisson Salazar[1], Evelyn Armendáriz[1], Kevin Álvarez[1], José Núñez[2], Mario García[1]

[1]Universidad Técnica de Ambato
[2] Universidad Técnica de Cotopaxi

johnpreyes@uta.edu.ec; darwinsaldas@uta.edu.ec;esalazar9560@uta.edu.ec; earmendariz0592@uta.edu.ec; kalvarez0229@uta.edu.ec; jose.nunez@utc.edu.ec mariogarcia@uta.edu.ec

Abstract. The scheduling of the operations of a manufacturing system is recognized for its efficiency in establishing a characteristic rate of production based on the forecasting of the ending date of an order. However, progressive planning focused on the footwear industries has not been studied in detail, since it is limited by the use of machines and supply according to the demand of the production line, whose development is based on just in time. The study proposes a finite progressive planning model in the area of footwear assembly that begins with analysis of the demand and identification of manufacturing constraints in order to establish an optimal ordering sequence. The results show manufacturing requirements through production orders that automatically determine production shifts per order, through experimentation of scenarios, the 25% increase in productivity indicators and a 31% improvement in efficiency are established. This improvement represents higher benefits for the industrial sector when establishing planning in the workplace.

1. Introduction

In this globalized world, companies have as priority to keep the customer satisfied, for this they seek to develop timely and quality products, trying to decrease production and storage costs, and that is where Fair Time Planning is born, i.e. JIT[1]. This implies that the manufacturing companies require radical changes to achieve the required quantity and quality of their productions and respond to market needs quickly, being essential a correct selection of the production planning and control system[2], especially for manufacturing plants that seek efficiency and effectiveness in their processes[3].

Programming production operations is known as Manufacturing Execution Systems (MES). A MES is an information system that schedules, dispatches, tracks, and controls the production of the plant. Production scheduling systems can be of infinite or finite load, infinite load occurs when the work is assigned to a work center according to what is needed over time and in a finite load approach is programmed Detail every resource in the preparation and run times for each order[4]. Developing a planning based on a JIT system can be considered a habitual problem, since engineers often do not have the time or the appropriate tools to make good production management[5], for this time dimension, the most common is progressive programming. Given that the footwear industry does not require very high production volumes[6], it is essential to consider the application of a progressive program that is developed under the JIT methodology, since any quantity that exceeds minimum required is considered a waste by investing effort and material in something that is not needed at that time[4].

By using a Just In Time system, it creates an advantage against competition and allows the entrepreneur to be more productive facilitating the response of the company to the needs of the demand, fulfilling the commitments acquired with the customer, delivering the products in the shortest possible time, thus increasing reliability in delivery,maximizing the new money generated by the company (throughput) and offering an optimal solution to small footwear companies[7].

Referring to the above, the Finite Progressive Planning, part from the need of companies to use optimization techniques to improve processes. Nowadays, many industries optimize process based on mixed-integer programming considering logistics at the same time [8]. This kind of programming presents obstacles since it exclusively depends on the effectiveness of the inventory management. These obstacles or issues are the lack of participation of workers and resistance to continuous improvement, cultural differences, high inventories levels, long term, market volatility, forecasting errors [9]. In the Indian industry, this programming is analyzed using performance indicators, such as cost reduction, productivity, reduction in the delivery time, variety of products, response and employee engagement [10].

Also, for other studies in this subject the uncertain random multilevel programming is proposed for problems involving production control [11].

Over the years, JIT production proposals have been developed whose approach takes place in such a way that all the company's capabilities can be aligned in a single production system, without neglecting the restrictions that are presented in such a system, given that effective management of these constraints is critical to reducing the uncertainties of a process and improving workflow, however, current approaches to the removal of constraints are fragmented and largely depart from human commitments[12], the timely control of these resources allows us to maximize resources and to have a minimum and adequate rotation of inventories, which is the objective of all cost systems[13], since it is a strategy that allows to improve the performance of companies, since all seek a continuous improvement controlling several aspects of the productive process in a collaborative way from the same supply chain until the control of the restrictions in the process[14], which through joint actions succeed.

One of the tools closely related to the JIT methodology refers to the pull production systems, whose objective is to let demand orient production and begin to manufacture the product after demand has reached the system[15]. This manufacturing principle allows companies to develop a program of operations that ensures that short- and medium-term decision-making is efficient, so that the analysis of constraints is essential in the optimization of results [16]. Hence, the need arises to consider a potential problem within production scheduling, which focuses on identifying and exploiting process constraints, so that organizational objectives can be achieved, as well as increased productivity and Competitiveness, improving costs and positioning the company in the market[17].

As a complement to the scheduling of operations, the scheduling of production is considered as a method to monitor the progress of a job and the load it produces on the manufacturing cell, maintaining as a limit the date of completion or delivery of the product Offered[18].

Studying and applying methods that allow scheduling production through Lean tools, reduces waste of time, work in process and raw materials, and can be projected in the future with large-scale studies of sustainable production as in the case of countries in Asia using appropriate sustainable business models and green supply chain management [19], this would imply the redesign of consumption patterns, which should not imply a decrease in the quality of life [20].The idea of a sustainable production and consumption is becoming a widely accepted social goal around the world [21].

The objective of this article is to show the experimentation of a model for progressive programming for the conventional footwear assembly area considering as demand variables and coordination of operations to produce weekly production schedules with a uniform workload. This influences the decision making about the use of resources and the pace of work conditioned by the delivery times of the finished product.

2. Methodology

This work uses an experimental methodology that describes a case study in the footwear sector regarding a progressive production system centered on the assembly area. The investigation frames its development in several leather shoe manufacturing organizations of the province of Tungurahua - Ecuador, from which a total of 32 footwear lines are obtained, of which a sample of 3 different models of the most produced in the Conventional assembly line the same as they are; Casual man, sporty woman and industrial safety. Each of these models represents different manufacturing lines.

The study begins with the analysis and identification of the activities and restrictions that take place in the assembly area of the companies, to establish a production of several models in the same working day, so as to avoid unnecessary stoppage of the process. Next, the sales forecast is generated, which allows a monthly and weekly production master plan to be applied, which facilitates the distribution of customer requirements through the demand curve, which describes the quantity of orders requested per order followed The actual production capacity is calculated, which interferes with the manufacturing time of each order and the quantity of human resources required for the manufacturing line, generating sequencing of production orders using a daily schedule that allows greater control over The reception times of sub-assemblies and delivery of finished product. Based on the information obtained, the calculation of indicators of productivity and efficiency of the system is developed by using it on two different production scenarios, one current and one proposed.

This research focused on the assembly area proposes a model of planning and sequencing of orders focused on the parameters and conditions of production of this process in a way that maximize the resources and limitations. The proposed model requires to develop a series of steps described below: Applying Restriction Theory (TOC) At this point, 3 of the 5 steps in TOC are developed, starting with the determination of all the restricted resources and bottlenecks in the assembly area, identifying the processes whose capacity is less than the required demand, and then deciding how to exploit the Constraints and subordinate the process to them[22]. Determining Sales Forecasting For the development of the forecast, the Winter Method[23] was used, where the systematic component of observed demand (Dt) has an estimate of the trend at the end of the period (Tt) and an estimated seasonal factor) And three smoothing constants for the level (α), trend (β) and for the seasonal factor (γ) with values ranging from 0 to 1. The forecast for future periods is given by:

$$Ft = (Lt - I + Tt - I)St \quad (1)$$

Observing the demand for period + 1, the estimates for the level, trend and seasonal factors are revised as follows:

$$Lt = \alpha(Dt + I * St) + (1 - \alpha)(Lt - I + Tt - I) \quad (2)$$

$$Tt = \beta(Lt + Lt - I) + (1 - \beta)Tt - I \quad (3)$$

$$St + p + I = \gamma(Dt + I)(Lt + I) + (1 - \gamma)St + I \quad (4)$$

Use of demand curves by footwear model the demand for footwear is concentrated according to the frequency distribution of the anatomical dimensions of the foot of the population according to the region that uses it, so it is concentrated differently in each size of footwear. By calculating and applying the demand curves by model, the quantity to be produced considering its individual demand for size is determined by applying Pull production concepts. This minimizes the inventories in process, final product and lost sales of the companies to maximize the availability of demanded products, as mentioned in Little Law proposed by Hopp and Spearman[24].

$$TH = WIP / CT \tag{5}$$

Where TH represents the Trougputh (Production Rate), WIP is the Inventory in process and TC refers to the cycle time. Develop the calculation of the weighted capacity considering the mix of models.

To develop a synchronized production in combination of models it is not enough to know the production capacity of each one, since it is necessary to obtain a real average value of production to schedule each shift of work of an efficient way in the whole installation. This value is defined as weighted production capacity, which depends directly on the number of models and order size of each. As a result, it has a real capacity that changes dynamically and adjusted to the reality of the constraints of the production plant.

Production Order Sequencing finally, to have a progressive production requires in footwear manufacture to have an inventory of molds that guarantees the constant flow of materials between all the processes in the production line and that this does not stop in some instant of time. At this point the production shifts are distributed for each model to be manufactured, considering the production limitation given the number of forms per model available for the process. Once defined the relevant points of the model proposed for the progressive programming of footwear is analyzed in detail each tool used, in order to establish the role they take within the hypothesis approach of the study and its contribution in the verification.

3. Results

Tungurahua, one of the main shoe producing provinces in Ecuador[25], offers a diversified set of shoe models, taking into account that most follow the same pattern in their production processes. This begins with the punching, where the pieces required for the model are cut, then sent to the trimming process which is responsible for making a subassembly by sewing the die pieces. In the assembly process, the trimmed cut takes the form of the foot by means of shoe molds (shapes) to later join the base of the product with the sole.

3.1. Identification of Constraints

By observing the productive process of the assembly area, two main resources are identified that limit the fulfillment of the requested orders, the first of them focuses on the quantity of shoes that exist by model, which given the small amount required Apply a combined production so that the production line is not stopped because of a lack of material resources and demand is satisfied without any inconvenience.

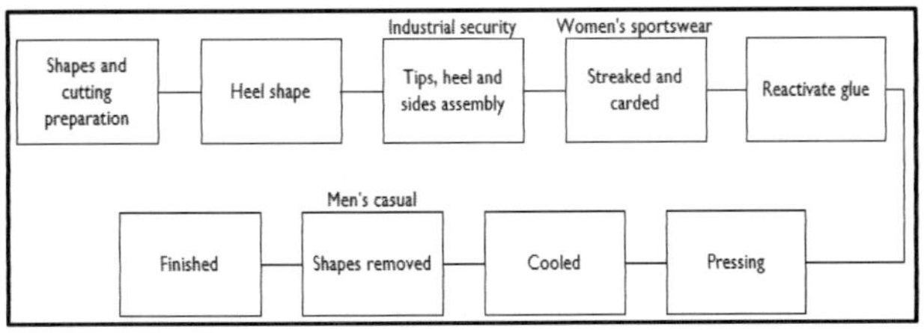

Figure 1. Capacity restricted by footwear model.

The second limitation focuses on the restricted capacity of the process, because not making decisions based on this resource can overload the bottleneck stopping production and causing unnecessary costs. Figure 1 shows the footwear assembly manufacturing sequence, which remains constant for the 3 models analyzed; However, each model has a different bottleneck that is used to develop production scheduling, thus, the restricted process of the casual model, the final drawing, For the sporty it is the process of scratching and grinding and the safety is limited by the assembly of points.

3.2. Demand Analysis

The analysis of the sales made in the last years comprises a fundamental step for the investigation, because it is important to know the evolutionary evolution that has been registered of the models of footwear under study, with the aim of predicting the sales that could be generated in the medium term. Through the application of a forecast based on time series and due to the periodic fluctuations that are influenced by factors it is determined that the affectations of the past and that continue to disturb future time series.

Figure 2 shows the sales history of each model studied and the forecast calculated by framing in a box, from the information obtained the Winter time series method is applied in order to obtain the trend with adjusted seasonal values Of the group of models and considering the values of the constants α = 0,1; B = 0.1; Y = 0.1 corresponding to the revised estimate of the initial level, trend and seasonal factor respectively. Based on the straight line obtained, it is established that the safety model exceeds the base demand of pairs of footwear, while the sport model maintains a margin of variation closer to the tension line and the casual model presents a reduction the trend calculated.

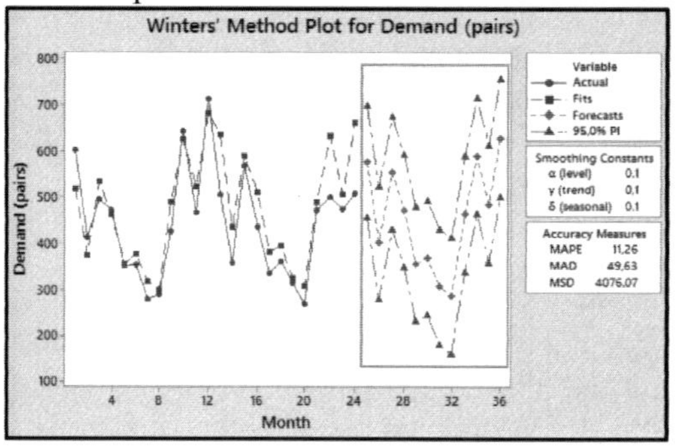

Figure 2. Sales history and forecast in Minitab software 17.

The main purpose of the analysis of the demand curve is to automatically represent the quantities required for each model by applying the percentage of sales sold in previous years. Figure 3 shows the demand curve formed by the points of intersection between two variables: model sizes and production requirements.

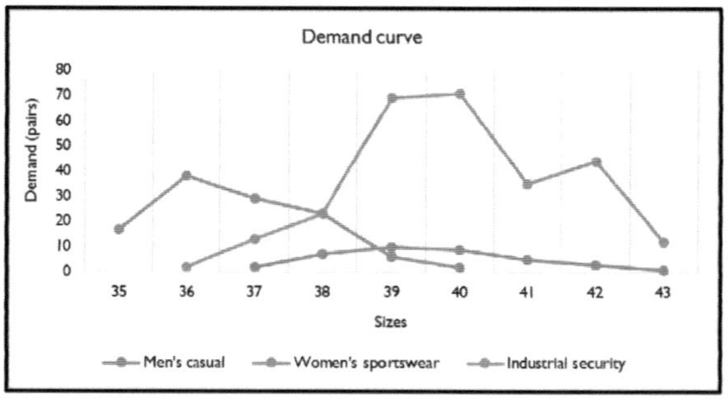

Figure 3. Demand curve by model studied.

Each model adopts a different curve, since the production requirements vary according to the market segment, whereas for the sports model there is a concentration between sizes 36 and 37; For the safety model the focus is on sizes 39 and 40; While the casual model presents an almost linear trend of production since the sizes 39 and 40 have a slight concentration with respect to the rest of sizes.

3.3. Aggregate Production Planning

For the assembly process, we have traditionally considered a series of aggregate production plans based on overtime, subcontracting, contracts and dismissals whose purpose is to adjust to the conditions in which the assembly process takes place; However, it is proposed to calculate the number of workers for the manufacturing cell based on the percentages of utilization of equipment and machinery, as shown in Figure 4. The results show a need for 12 workers obtained based on to the sum of the capacities.

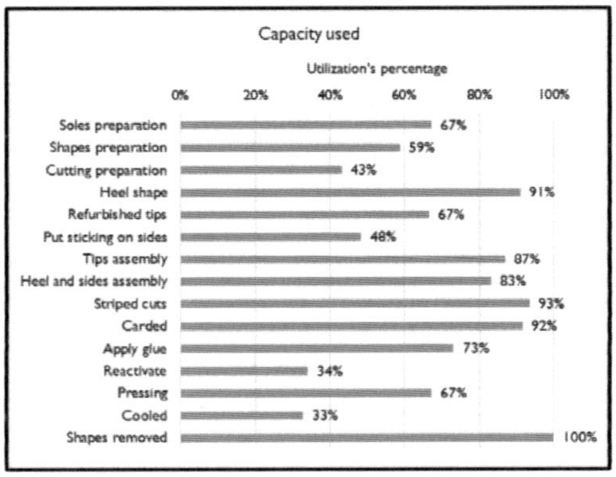

Figure 4. Utilization of machinery capacity

The proposed aggregate plan represents a variation of those already known and consists of producing on-demand with progressive programming, retaining the number of workers calculated and employing overtime when demand can't be satisfied with normal production capacity. In the case presented, the aggregate plan generates a saving of 16.06% over the traditional least-cost aggregate plan, which translates into a considerable direct cost reduction for the footwear producing sector. The proposed aggregate planning is characterized by distributing the high demands to the months in which the quantities required are less than the production capacity; While in some cases the use and coordination of ex-hours is required to meet the demand for future months, which avoids the increase of the labor load in the months with the highest demand and balances the production capacity to take advantage of the company resources.

3.4. Obtaining Production Orders

When considering a progressive production system, it is established that assembly sends the production requirements to the trimming and sub-assembling sub-assemblies through production orders that allow to synchronize and control the flow of work processes and loads. For the case under study, demand is calculated and calculated using demand curves, see Table 1.

Table1. Production Order.

MONTH 2													
WEEK 2													
CODE	MODEL	SIZES											TOTAL PRODUCTION (pairs)
		34	35	36	37	38	39	40	41	42	43	44	
CA873	MEN'S CASUAL				2	7	10	9	5	3	1		37
SK568	WOMEN'S SPORTSWEAR		17	38	29	23	6	2					115
S09	INDUSTRIAL SECURITY			2	13	23	69	71	35	44	12		269

3.5. Work shifts Calculation

The definition of work shifts involves the use of a couple of tools; The first so-called average monthly production capacity whose purpose is to allocate quantities to be drawn based on the relative importance of the models to each other. The second tool based on the calculation of the number of shifts required per order, for which we proceed to apply equation (1) on each model, establishing a relation between the weekly production capacity and the demand demanded in the same week Number of Shifts = Demand / (Weekly Capacity) (6) Since the value generated by equation (1) is obtained in weeks, there is a need to convert that time into days, as this facilitates the understanding and assignment of work schedules. The weighted production capacity is calculated by applying the Microsoft Excel "Search Target" tool with which the values of the actual capacity of each model are automatically distributed according to the standard and their quantity demanded by production order, which varies according to the Work shifts. This value is dynamic and changes at each shift, adjusting to demand.

3.6. Order Sequencing

Order sequencing allows the definition of the workload according to the production capacity per shift and the modeling scheduling according to the priority rules defined on the basis of the material constraints identified (number of shims of shoes by size and model). To start with the sequencing, it is necessary to elaborate a schedule of activities that establish the start and end date of the assembly sub-assemble by using Gantt diagrams to-command information of the production times of punching and trimming obtained in research independent.Table 2 gives the start of the production of footwear with the punching process, which takes a total of 6.2 hours to meet the demand established in the production order, followed by the trimming area with a time of 13.6 hours and Assembly with 19.8 hours. The example cited corresponds to an ordinary week of 5 days, whose production order is displayed in section 3.5.

Table2. Gantt diagram of production order footwear.

Process	Production hours	Monday	Tuesday	Wednesday	Thursday	Friday
Cutting	6,2					
Trimmed	13,6					
Assembly	19,8					

Figure 5 shows the sequencing of the models and sizes to be produced according to the order of production, taking into account that the days used correspond to those described in Table 2 by means of the Gantt chart. For this scenario the production time is two and a half days, starting production with a total of 80 pairs of shoes of the safety model combining several sizes due to the missing shoe sizes. The same rule of priority applies in the following day in which 49 pairs of the sports model are produced, followed by 13 pairs of casual and 108 pairs of safety footwear.

The scheduling of jobs for the assembly area depends directly on the number of shims per size and model so that those models with high demands complete several work cycles per shift, taking into account that the combination of models ensures the Return of logs to the beginning of the cellallowing to continue fulfilling the entered order, as indicated in the last day of work of Figure 5.

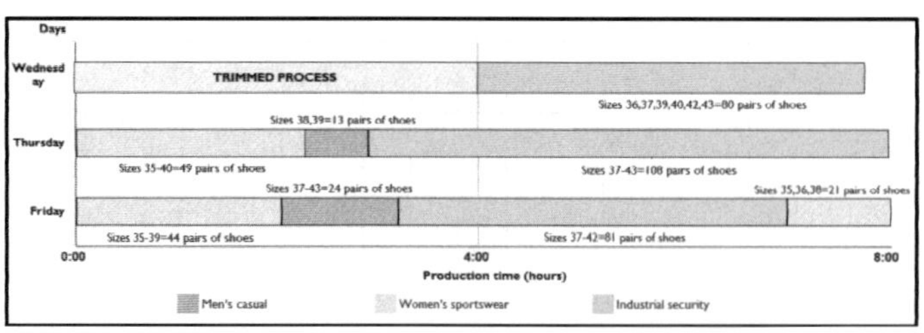

Figure 5. Sequencing production orders footwear.

4. Discussion

The analysis of two production scenarios, taking into account the variations in demand, current and proposed resources of companies, validates the optimization of the production system through the use of indicators of productivity and efficiency of the assembly area.

Figure 6 shows a comparison of production efficiency, for this it is established that the blue curve represents the current monthly capacity whose base is defined with 160 pairs / shift while the red curve represents the calculated monthly capacity that varies according to the demand per model required monthly.

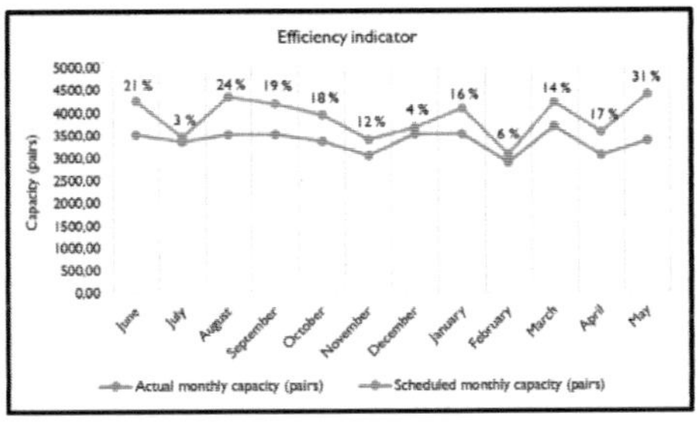

Figure 6. Efficiency variation current and proposed system.

Around the data shown in Figure 6, it is established that the research gives companies the possibility of improving their production by 3% and 31% through optimization of resource management and sequencing of activities. In addition, the calculation of partial productivity is developed, since the project does not consider a cost analysis, which evaluates the degree of utilization of the number of man-hours that affect the footwear assembly process. Figure 7 shows the productivity variation in a first scenario of demand related to the human resource that companies have. The initial number of workers was 15 and the pro- posed using the programming model were 12, resulting in an increase in productivity of this first scenario corresponding to 25% of current production.

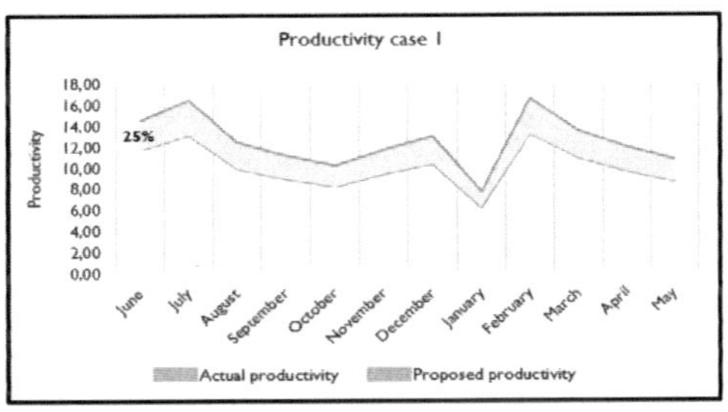

Figure 7. First stage productivity variation.

Under the consideration of the same human resources used for the previous case, a second production scenario is developed based on the forecast data obtained; Through this information is generated Figure 8, which also shows an increase in productivity that corresponds to 25% with respect to the current system of the company. Based on the two scenarios shown, it is established that for both cases the productivity is maintained a growth of 25%, a fact that undoubtedly creates a competitive advantage of the organizations compared to their competitors by reducing manufacturing costs in direct labor.

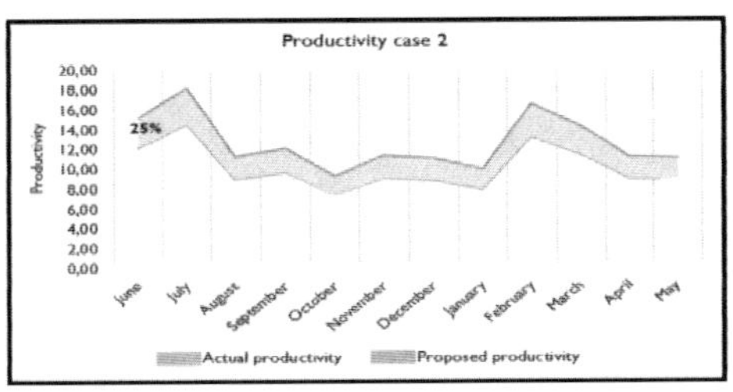

Figure 8. Productivity variation according to scenario.

5. Conclusions

The development of the daily sequencing of production that is part of the study influences the use of the working day and the use of material resources in each work cycle, which makes the progressive programming model proposed as an innovative tool for optimization of Operations for the conventional footwear assembly area by influencing the direct reduction of manufacturing times. One of the most important aspects of the programming model focuses on the application of OCD that alters its development perspective, taking into account the limited amount of resources available by shoe model leading to produce based on a mix of models, in order to avoid increases in production times or unnecessary stoppages of the production system. The planned planning model brings with it the maximization of benefits to footwear companies by increasing their production levels, represented by efficiency indicators that reach 31% of the current production standard of companies in the industrial sector. This fact supports the importance of opting for the proposed model as a method for improvement and decision making in the process. The application of the production planning and scheduling model satisfies the material flow needs of the footwear assembly process, since it is a novel method based on the proper use of JIT methodologies. The efficient allocation of human resources provides the opportunity to reduce direct manufacturing costs, maintaining a constant increase in productivity, taking a value of 25% in relation to the current productivity of companies. It is recommended that footwear companies opt for the use of the proposed product planning and sequencing model, since it represents a tool that adapts to the conditions of conventional assembly and allows the potential improvement of productivity through the program-progression of each footwear model considering the demand aspects and production constraints.

Acknowledgments

The authors thank the National Footwear Chamber of Ecuador and the Technical University of Ambato for the support provided during the execution of this work within the framework of the research project called "Operational optimization based on a lean dynamic system of alert of failures in the production processes for the footwear industry".

References

[1] F. Elizalde, Y. Silva and Y. Ríos, "Planeación justo a tiempo: Soluciones óptimas mediante reformulaciones convexas, Just-in-time planning: Optimal solutions through convex reformulations," *Revista Ingenierias,* vol. 14, no. 51, pp. 43-50, 2014.

[2] A. Tamayo García and I. Urquiola García, "Design of a Process for Planning and Controlling Production by Using Mathematical Tools," *Revista de Métodos Cuantitativos para la Economía y la Empresa,* vol. 18, pp. 130-145, 2016.

[3] M. Arango Serna, L. Campuzano Zapata and J. Zapata Cortes, "Manufacturing process improvement using the Kanban," *Revista ingenierías Universidad de Medellín,* no. 14, pp. 221-233, 2015.

[4] R. Chase y R. Jacobs, Operations Management. Production and supply chain, Mc Graw Hill Education, 2014.

[5] R. Solar, I. Chacón and M. Ponce, "Aggregated production plan in sawnwood mill. Case of study for small industry," *Maderas. Ciencia y Tecnología,* vol. 10, no. 2, pp. 77-92, 2008.

[6] J. Reyes, K. Alvarez and R. Vasquez, "Dynamic buffer management for raw material supply in the footwear industry," *Journal of Industrial and Intelligent Information,* vol. 4, no. 1, pp. 1-8, 2016.

[7] V. Ortiz and A. Caicedo, "Optimal production scheduling in a small shoe business in Colombia," *Revista de Ingeniería Industrial,* vol. 35, no. 2, 2014.

[8] C. Damião Rocco y . R. Morabito, «Scheduling of production and logistics operations of steam production systems in food industries: a case study of the tomato processing industry,» *Journal of*

the Operational Research Society, vol. 65, p. 1896–1904, 2014.

[9] J. Jadhav, S. Mantha y S. Rane, «Barriers for successful implementation of JIT: a manufacturer perspective,» *International Journal of Procurement Management,* vol. 7, nº 3, pp. 316-342, 2014.

[10] D. Tewari, A. Singh y P. C. Tewari, «Ranking of Performance Indicators,» *IUP Journal of Operations Management; Hyderabad,* vol. 1, nº 15, pp. 25-34, 2016.

[11] H. Ke, T. Su y Y. Ni, «Uncertain random multilevel programming with application to production control problem,» *Soft Computing,* vol. 19, nº 6, p. 1739–1746, 2015.

[12] J. Wang, W. Shou, X. Wan and P. Wu, "Developing and evaluating a framework of total constraint management for improving workflow in liquefied natural gas construction," *Construction Management and Economics,* vol. 34, no. 12, pp. 859-874, 2016.

[13] L. Mendoza and M. Vega, "Theory of constraints and continuous improvement process vs just in time methodology (JIT) and ABC cost," *Dictamen libre,* vol. 14, pp. 7-13, 2014.

[14] J. Mayer, M. Borchardt and G. M. Pereira, "Methodology for the collaboration in supply chains with a focus on continuous improvement," *Ingeniería e Investigación,* vol. 36, no. 2, pp. 51-59, 2016.

[15] A. Mora, J. Tobar and J. Soto, "Comparison and analysis of some production control systems type pull, using simulation," *Revista Scientia et Technica,* vol. 51, pp. 100-106, 2012.

[16] G. Ramírez, M. Torné y J. Orejuela, «Scheduling hoppers filling operations in the animal concentrated feeding industry,» *Revista Ingenierías Universidad de Medellín,* vol. 11, nº 20, 2012.

[17] G. Villagómez, J. Viteri y A. Medina, «Teoría de restricciones para procesos de manufactura,» *Revista Enfoque UTE,* vol. 3, pp. 14-28, 2012.

[18] L. Krajewsky, L. Ritzman and M. Malhotra, Operations management: process and value chains (11th edition ed.), Pearson, 2015.

[19] Ming-Lang Tseng , Shun Fung Chiu , Raymond y A. B. Siriban-Manalang , «Sustainable consumption and production for Asia: sustainability through green design and practice,» *Journal of Cleaner Production,* vol. 40, pp. 1-5, 2013.

[20] J. H. Spangenberg, «Design for Sustainability (DfS): Interface of Sustainable Production and Consumption,» *Handbook of Sustainable Engineering,* pp. 575-595, 2013.

[21] A. Azapagic, L. Stamford, L. Youds y C. Barteczko-Hibbert, «Towards sustainable production and consumption: A novel DEcision-Support Framework IntegRating Economic, Environmental and Social Sustainability (DESIRES),» *Computers & Chemical Engineering,* vol. 91, nº 4, pp. 93-103, 2016.

[22] E. Goldratt, La Meta (3 ed.), Mexico: Ediciones Granica, 2008.

[23] R. H. Ballou, Business logistics/supply chain management: planning, organizing, and controlling the supply chain, India: Pearson Education, 2007.

[24] W. Hopp and M. Spearman, Factory Physics (Third Edition ed.), Waveland Press, 2011.

[25] Mipro, "Mipro," [Online]. Available: http://www.industrias.gob.ec/bp131-ficce-2015-oportunidad-para-conocer-a-la-industria-del-calzado-ecuatoriano-con-calidad/. [Accessed 1 February 2017].

[26] M. Fazel Zarandi, M. Khorshidian y M. Akbarpour Shirazi, «A constraint programming model for the scheduling of JIT cross-docking systems with preemption,» *Journal of Intelligent Manufacturing,* vol. 27, nº 2, p. 297–313, 2016.

Prediction of Service Life for Assembly with Time-variant Deviation

Wenhui Zeng, Yunqing Rao, Chenxi Long,Peng Wang,

The State Key Laboratory of Digital Manufacturing Equipment and Technology, School of Mechanical Science & Engineering of Huazhong University of Science and Technology, Wuhan 430074, PR China

raoyqhust@gmail.com (Yunqing Rao), zengwenhui3242@gmail.com (Wenhui Zeng)

Abstract. During operation, the time-variant deviations, such as deformation, thermal expansion, friction and wear always occur and affect the mechanical performance and service life of assembly. In this paper, a methodology for the prediction of service life for assembly with time-variant deviations is proposed. Firstly, based on the modified Unified Jacobian-Torsor model and Monte Carlo simulation, according to the distribution the geometric and dimension tolerances are randomly generated to limit the variations of surface of part. Secondly, the deformations caused by load and thermal expansion are obtained by finite element analysis, and considering the friction and wear, the prediction model of service life is constructed subject to the constraints. At last, an application is given to illustrate the prediction model, and the influences of time-variant deviations on the assembly deviation and service life are analyzed.

1. Introduction

Tolerance analysis is widely used in the management of dimensional and geometrical variations of products for engineers. These variations from the nominal dimensions are stochastic and can be described by their statistical distribution, and their assembly variations can be calculated by many tolerance analysis models, mostly with the aid of Monte Carlo simulation [1-3]. These traditional analysis models rely upon rigid body motion to model the propagation and accumulation of variations that occur in an assembly of parts. However, during the operation, it would be more realistic that the components of assembly are flexible and usually have the time-variant deviation due to inertia effect (gravity, angular velocity, angular acceleration, etc.), temperature effect and friction. In addition, with the mobility of parts, the wear between contact surfaces always exists after a few of times whether they are rigid or flexible. If ignoring these deformations and wear, it could lead to an inaccurate assembly deviation and further to affect the performance, reliability and service life of products. In this paper, a methodology is proposed to predict the service life of product caused by time-variant deviation due to the preload of bolts, centrifugal force, thermal expansion and friction.

In recently years, a few studies focus on the tolerance synthesis with considering the time-variant deviation. The thermo-mechanical strains are integrated into the tolerance analysis model to control the clearances at the tip of a high pressure turbine blade with different operating regimes of a helicopter engine [4]. A variational method of tolerance analysis using a multi-physical approach was proposed by Pierre, Teissandier[5] to validate geometric specifications, contact specifications and thermo- mechanical specifications. An approach proposed by Benichou and Anselmetti [6] considered thermal expansion of parts integrated within functional tolerance. Based on stochastic simulations, Grandjean, Ledoux [7] analyzed the influence of form defects in assemblies subject to local deformations and mechanical loads and concluded that even if the form defects are limited, they can lead to a non-compliant assembly. As the local deformations of parts occur, the cumulative effects of component parameter drift and tolerance variation many eventually cause the output variables of interest to deviate from the design specifications. Then after deviation propagation and accumulation along the kinematic chains, the final assembly deviation would beyond the functional requirement during service life. Therefore, it would be more realistic that it could obtain a more accurate calculation result of assembly deviation with considering the time-variant deviation according to the works by Mazur [8, 9].

In this paper, based on the modified Unified Jacobian-Torsor model and Monte Carlo simulation, according to the distribution the geometric and dimension tolerances are randomly generated to limit the variations of surface of part. Secondly, the deformations caused by load and thermal expansion are obtained by finite element analysis. Considering the friction and wear, all of these deformations and wears are integrated into the tolerance analysis model to construct the service life model, and finally the influences of time-variant deviations on the assembly deviation and service life are analyzed.

The remainder of this paper is organized as follows. Section 2 gives the problem description of assembly with time-variant deviation. Section 3 gives the tolerance analysis for assembly with time-variant deviation. Section 4 gives the prediction model of service life of assembly. An application is given in section 5 to illustrate the approach of this paper and section 6 gives the discussion and conclusion.

2. Problem description

In this paper, the blade bearing of controllable pitch propeller is regarded as an application to illustrate the methodology of predicting the service life with time-variant deviation. Figure 1 is the blade bearing of the controllable pitch propeller (CPP) and Figure 1 (right) is the partial enlarged view of O-ring hydraulic seal. This assembly consists of blade foot, blade carrier, O-ring and hub. The blade carrier is mounted on the hole of hub. The O-ring is extruded to fill up the O-groove between the blade foot and hub; and it can form O-ring hydraulic seal to prevent the seawater flow into the hub and the hydraulic oils flow into the seawater. Simultaneously, the blade carrier and bladed foot are fastened with bolts and no relative displacement between them.

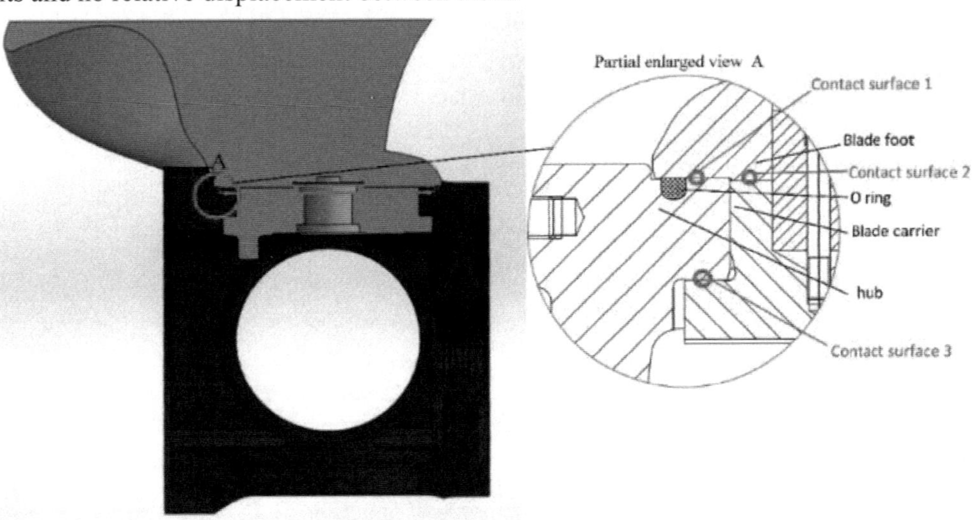

Figure 1. Blade bearing of the controllable pitch propeller.

Currently, a few researches focus on the blade bearing of controllable pitch propeller. The influence of sea state on fretting behavior of blade bearing in a CPP was described by Godjevac, Van Beek [10]. In fact, the failure of the blade bearing is mainly presented as leakage which is caused by the decreased compression ratio of O ring during the service life. During the actual service, when the wear of contact surface 3 increased, the gap (contact surface 1 in Fig. 1) between the blade foot and hub could increase, and it will lead to a rebound of O-ring. Then its compression ratio could be decreased and if it is less than the required compression ratio, it will result in leakage. However, if the initial gap is too small, the increased temperature of the hydraulic oil in the cavity of hub, bolt preload and centrifugal force may result in deformations of parts; these deformations may further reduce the gap, or result in interference fit between the blade foot and hub. Then it could be very hard to adjust the pitch of CPP, and the blade foot and blade carrier could be hard to swing when needs to change the pitch.

Therefore, in order to extend the service life, the initial gap should be small, but in order to ensure the flexibility of adjusting the pitch, the initial gap should be larger than the functional requirement. The gap during the service life is a time-varying variable, which consists of three parts. The first part regarded as d_{g-t} is formed by the dimension geometric tolerances; the second part regarded as d_{g-r} is formed by the deformations due to the centrifugal force, bolt preload and temperature; the third part regarded as d_{g-w} is caused by the friction and wear, and this part is related to the service life. Then the gap can be written as:

$$d_g = d_{g-t} + d_{g-r} + d_{g-w}(T) \tag{1}$$

where T is the service life of blade bearing. In this paper, these time-variant deviations are considered to predict the service life of the blade bearing, and the influences of these time-variant deviations on the assembly deviations are analyzed.

3. Tolerance analysis with time-variant deviation

3.1. Tolerance analysis with considering the geometric tolerance

A detailed mathematic deduction of the variations and constraints of planar feature specified by dimension, orientation and position tolerance has been proposed by Roy and Li [11]. In this paper, all the variations of related planes are simultaneously limited by dimension tolerance and geometric tolerance. Figure 2 (left) shows the related tolerances of parts of assembly, and the gap between the blade foot and hub is determined by the dimension tolerances X_3 of hub, X_2 of blade carrier and of blade foot as well as the their corresponding geometric tolerances t_1, t_2, t_3 and t_4. Figure 2 (right) shows the connection graph of this assembly.

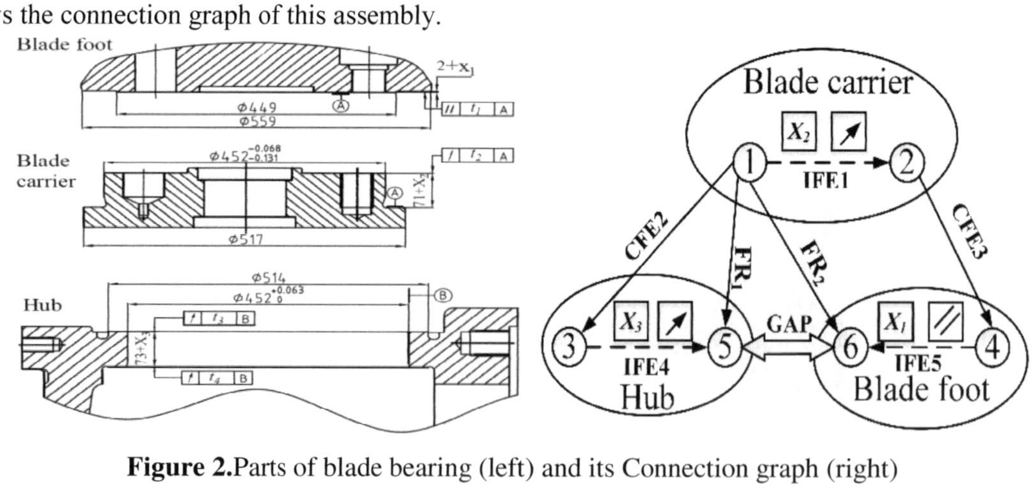

Figure 2. Parts of blade bearing (left) and its Connection graph (right)

In figure 2 (right), there are two FRs, and the gap is formed by the two functional requirements. The fit of blade foot and blade carrier is perfect, and CFE3 is null. With considering the geometric tolerances, the functional requirements can be calculated by the modified method of Unified Jacobian-Torsor model proposed by Chen, Jin [12]. The more detailed description about the modified Unified Jacobian-Torsor model has been introduced in [12]. Monte Carlo method is used to generate the satisfied random values of each torsor for the components according to variations and probability distributions based on the work of Ghie, Laperrière [13].

3.2. Finite element analysis

After constructing the 3D model of the blade bearing of CPP, the next step is to develop a finite element model to determine deformation of various components and their effects on the gap between the blade foot and hub. The material properties of various elements of the analysis model are listed in Table 1. The periodic boundary conditions are used to define the symmetry surface. Figure 3 shows the finite element analysis model of this assembly, and then the next step is to determine deformation due to the temperature, centrifugal force and preload of bolt.

Table1. The material properties of each parts of the analysis model

Material property	Hub	Blade foot	Blade	Blade carrier	Bolt
Material	CuNiAl	CuNiAl	CuNiAl	42CrMo4	Structural steel
Density (kg/m^3)	7530	7530	7530	7380	7850
Modulus of elasticity (MPa)	111	111	111	212	210
Poisson's ratio ()	0.32	0.32	0.32	0.3	0.3
Yield strength (MPa)	260	260	260	550	250
Tensile strength (MPa)	650	650	650	800	460
Thermal conductivity (W/mK)	41.9	41.9	41.9	44	60.5

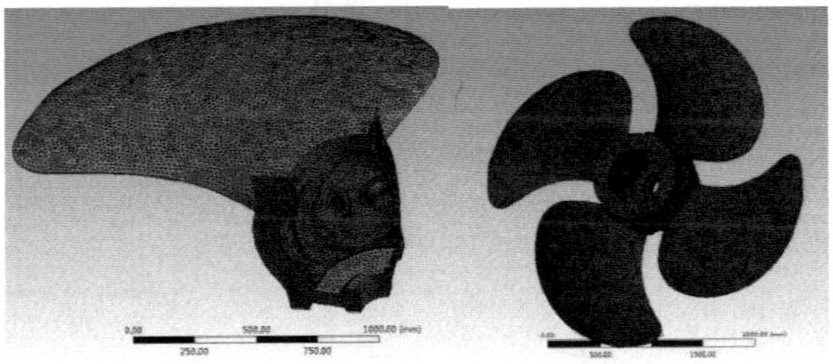

Figure 3. The finite element analysis model of the assembly

The first step is to establish the steady-stage thermal analysis. In this assembly, the temperature of all surfaces covered by the hydraulic oil is equal to the temperature of oil including the inner surfaces of hub, blade carrier and local surface of blade foot. The temperature of outer surfaces soaked in the sea water is 19 which are equal to the temperature of sea water. The alarm temperature of oil is 60 and the room temperature is 25 ° C, the variable range is 25-60 ° C. The results of steady-stage thermal analysis will be imported to steady-state structural analysis for next simulation.

The centrifugal force is generated by the rotation speed of blade. The range of rotation speed is 60r/min~180r/min according .The ranges of bolt pretension of the bolts are 320 KN~390KN.

According to characterization requirement and sampling frequency, the two mating planar surfaces of contact surface 1 were discretized as a mesh. The deformation of nodes in the mesh can be obtained from the finite analysis results. The data are represented as follows:

$$A_{di} = \begin{bmatrix} x_{i1} & y_{i1} & z_{i1} \\ x_{i2} & y_{i2} & z_{i2} \\ \vdots & \vdots & \vdots \\ x_{in} & y_{in} & z_{in} \end{bmatrix} \tag{2}$$

$$d_{g-r} = (A_{d1} - A_{d2}) \tag{3}$$

where i ($i = 1, 2$) is the number of mating surfaces, is the number of sampling points on each surface, the matrix comprised of finite element analysis data of the surface, x_{ij}, y_{ij} and z_{ij} are the coordinate values of the sampling point on the surface, respectively.

3.3. *Time-variant deviations of gap*

The wear of blade bearing of a CPP is caused by sliding motion and fretting motion. The wear rates of sliding motion and fretting motion both are related to the materials of blade carrier, and the related researches are given in [14].

In order to estimate the total wear of a CPP, it is necessary to know how much pitch is being used in real service conditions. When the total amount of pitch change is known, the total sliding distance in the blade bearing will be known for a certain time period. The total wear depth of blade bearings (h) can thus be calculated by combining the total sliding distance (S), specific wear rate (ω) and load (σ) into the following equation:

$$h = w \cdot S \cdot \sigma \tag{4}$$

The total sliding distance is related to the times n_s of total pitch change of one day, design pitch angle φ_d, the blade bearing radius r and the service life T, it can be written as:

$$S_{sliding} = n_s \cdot \frac{\varphi_d}{360} \cdot 2 \cdot \pi \cdot r \cdot T \tag{5}$$

The total fretting distance is related to the fretting amplitude d_{fret}, design revolutions per minute rpm and service life T, and it can be written as:

$$S_{fret} = 2 \cdot d_{fret} \cdot rpm \cdot T \tag{6}$$

Then the final wear depth can be written as:

$$d_{g-w}(T) = \omega_s \cdot S_{sliding} \cdot \sigma + \omega_f \cdot S_{fret} \cdot \sigma \tag{7}$$

4. Prediction model of service life

Service life of CPP is determined by the sealing performance of blade bearing and the flexibility of adjusting the pitch. Considering the statistical criteria, the tolerance intervals are the $\pm 3\sigma$ range, and the probability of eligible product is 99.73%. In order to ensure the flexibility of adjusting the pitch, the probability constraint in respect of the initial functional requirement can be presented as:

$$P_{grap} = P((d_{g-t} + d_{g-r}) > 0) \geq 99.73\% \tag{8}$$

The compression ratio range of rubber O-ring generally is 9%-15%, and the probability constraint of O-ring in respect of service life T can be written as:

$$P_T = P(0.09 \leq \frac{d_2 - ((d_{g-t} + d_{g-r})_{\max} + d_{g-w}(T) + x_4)}{d_2} \leq 0.15) \geq 99.73\% \tag{9}$$

Where d_2 is section diameter of O-ring in a free state; X_5 is the depth of O-groove. In this paper, two indicators are introduced to evaluate the performance of CPP, one is the probability $P_i(T_i)$ of service life T_i, and the other is the actual service life T_a. The prediction model of service life of CPP can be written as:

$$\text{Object: } f_1 = P_i(T_i), f_2 = T_a$$
$$s.t. \begin{cases} P_{gap} \geq 99.73\% \\ P_T \geq 99.73\% \end{cases} \tag{10}$$

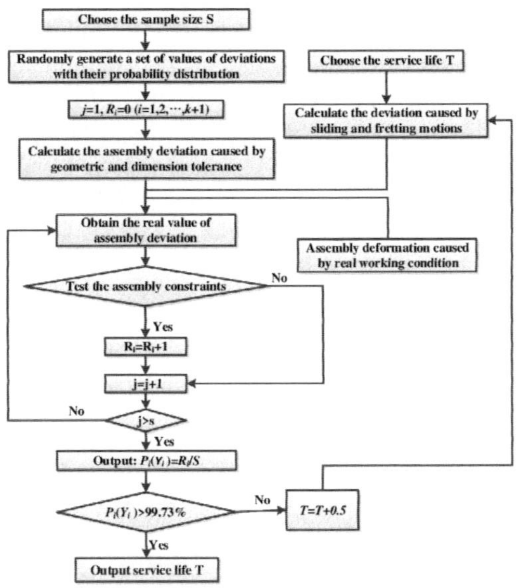

Figure 4. The procedure chart of prediction of service life of CPP

In this paper, an iterative method is used to predict the service life of blade bearing based on Monte Carlo simulation. The Monte Carlo simulation is used to calculate the probability of accepted assemblies. The procedure of prediction of service life of CPP is shown in Figure 4.

5. Experimental results

The related tolerance values of the blade bearing of CPP are given in Table 2. According to the Monte Carlo simulation and modified Unified Jacobian-Torsor model in [12], the first part of gap d_{g-t} is shown in Table 3 including the assembly deviations of FR_1 and FR_2.

Table2. Geometric tolerances and dimensional tolerances of each part

Tolerance	X_1	X_2	X_3	X_4	d_2	t_1	t_2	t_3	t_4
Value	2(-0.004,0.006)	71(0,0.03)	73(-0.09,-0.06)	15(-0.1,0)	17±0.15	0.02	0.02	0.02	0.02

Table3. Calculated results of FR_1 and FR_2

Functional requirement	Tolerance	Mean	Std	Var
FR_1	w_j	-0.075	0.0045	$2.024*10^{-5}$
FR_2	w_f	0.0160	0.00466	$2.174*10^{-5}$

5.1. Influence time-variant deviation on the assembly deviation

When the second part of gap d_{g-r} is considered, the maximum assembly gap between the hub and blade foot could increase and the minimum assembly gap between the hub and blade foot could decrease. Because, the deformation surfaces of parts are not uniform. The changes of the maximum assembly deviation and minimum assembly deviation caused by second part of gap d_{g-r} with the temperature, rotation speed and preload are shown in Figure 5, Figure 6.

In Figure 5 (left), the maximum assembly deviation increases with the temperature, but the minimum assembly deviation decreases with temperature. In Figure 5 (right), only the maximum assembly deviation increases with the rotation speed, and the minimum assembly deviation increases little when the rotation speed increases. However, in Figure 6, both the maximum and minimum assembly deviation changes little when the preload increases. Therefore, the temperature has the greatest impact on the assembly deviation, and effect both the flexibility of adjusting the pitch and the sealing performance of blade bearing. The rotation speed only effects the maximum assembly deviation, and further to effect the sealing performance, however, the preload has little influence on the assembly deviation.

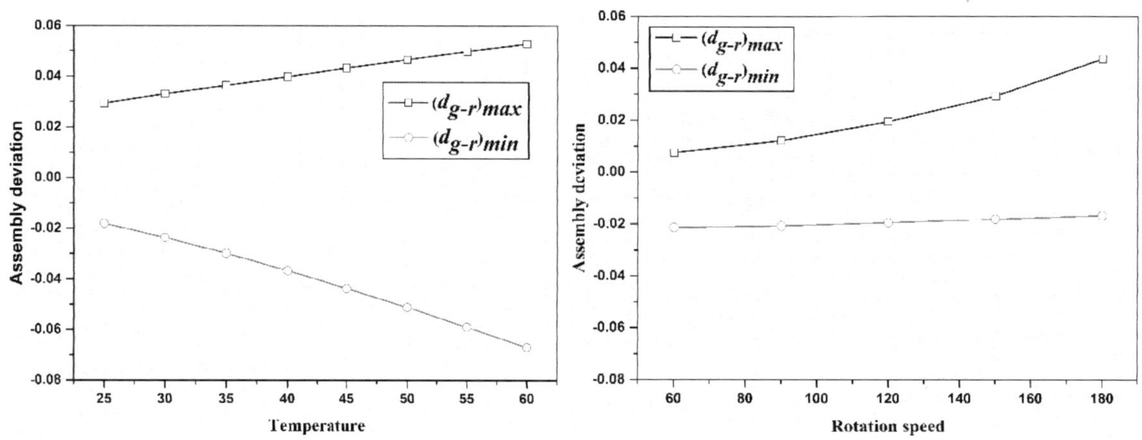

Figure 5. Second part of gap changes with the temperature (left) and rotation speed (right)

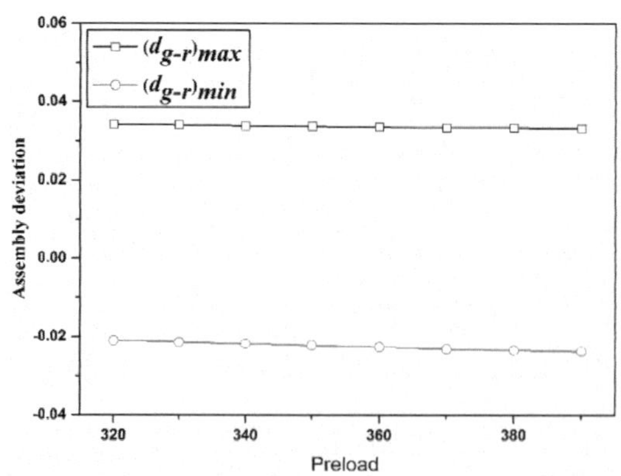

Figure 6. Second part of gap changes with the preload

5.2. Service life of blade bearing

Generally, the service life of products is related to the working conditions and operating frequency. The fretting behavior of blade bearing is effected by sea state and its frequency and amplitude to adjust the pitch[14], and the wear rate of the blade bearing is 0.54E-7 and the maximum load is 24N/mm^2.

In order to illustrate the methodology of predicting the service life of blade bearing, a ferry from Dover to Calais is used as an application to calculate the total amount of pitch change during one day which is introduced by Godjevac [14]. The sampling rate is 10Hz and the fretting amplitude (dfret) is assumed as 300μm. One voyage time (tvoy) is approximately 100 minutes and the total amount of pitch changes is 9.54 times of the design pitch angle (d_j =30deg). The total sliding distance is obtained by multiplying the design pitch angle with the blade bearing radius (r = 0.226 m). The rated rotation speed is 150r/min. This ship makes around 13 voyages per day (nvoy) and operates approximately 340 days per year (nday).

According to the equations (5) and (6), the sliding distance and fretting distance of one year can be obtained that:

$$S_{total} = 340 \cdot 13(9.54 \cdot \frac{30}{360} \cdot 2 \cdot 3.14 \cdot 0.226 + 2 \cdot 0.3 \cdot 10^{-3} \cdot 150 \cdot 100) = 44.772km \tag{11}$$

Then, according to the equation (7), the final wear depth of one year can be calculated:

$$H = 0.54 \times 10^{-7} \times 24 \times S_{total} = 58.02 \mu m \tag{12}$$

Since, the preload has little influence on the assembly deviation, and in this subsection, its influence on the service life is not analyzed. Table 4 shows the probability $P_i(T_i)$ of service life T_i and the actual service life T_a with different temperature. In Table 4, the actual service life T_a decreases when the temperature increases as well as the probability $P_i(T_i)$ of 5 years' service life. In Figure 7 (left), they are the probability distribution curves of service life T_i with different temperatures, and the normal curve is the probability distribution of service life T_i without considering the real working condition. It shows that, when the temperature increases, the probability distribution curve of service life moves left, and the actual service life T_a decreases, and finally the service life is reduced almost by 8% from 25℃ to 60℃.

Table4. Service-life changes with temperature (other parameters: rpm 150r/min, preload 390KN)

Temperature (℃)	60	55	50	45	40	35	30	25
of 5-year' service life	0.9377	0.9456	0.9543	0.96	0.9657	0.9713	0.9758	0.9801
Actual service life	4.13	4.18	4.23	4.285	4.345	4.4	4.455	4.51

Table5. Sliding and fretting distance with different rpm, and the total wear depth

Rotation speed（Unit: r/min）	60	90	120	150	180
Total sliding and fretting distance (one year) (Unit: m)	20898	28854	36809	44772	52723
Total wear depth (one year) (Unit: um)	27	37.4	47.7	58.02	68.33

Table6. Service life changes with rpm (other parameters: temperature 30℃, preload 390KN)

Rotation speed（Unit: r/min）	60	90	120	150	180
of 5-year' service life	100%	100%	100%	0.9794	0.1899
Actual service life	11.2	7.5	5.6	4.25	3.4

According to the equation (6), the fretting distance is related to the rotation speed. Different rotation speed has different total wear depth, and they are given in Table 5. Table 6 is the calculated service life with different rotation speed. Figure 7 (right) is the probability distribution curve of service life T_i with different rotation speed.

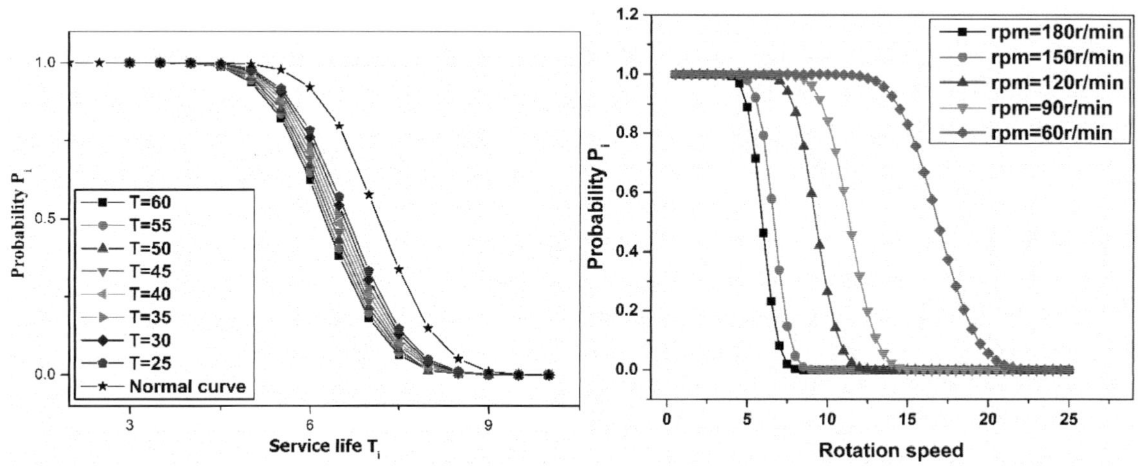

Figure 7. Probability distribution of service life T_i changes with temperature (left) and rpm (right)

From the Table 6 and Figure 7 (right), it can be seen that if the rated rotation speed increases, the actual service life decreases largely. When the rated rotation speed is set as 60r/min, its actual service life is 11.2 years; however, when the rated rotation speed is set as 180r/min, its actual service life is 3.4 years, almost threefold. Therefore, the temperature and rotation speed all effect the service life of the blade bearing of CPP, and if the temperature increased, the final service life would decrease by almost 8%; however, if the rated rotation speed were too large, the service life would be three times less. In actual service process, it is quite important to control the temperature and choose a reasonable rated rotation speed to extend the service life of blade bearing of CPP.

6. Conclusion

In this paper, a methodology for the prediction of service life for assembly with time-variant deviations is proposed, and the tolerance analysis model with considering the time-variant deviation is constructed. The assembly deviation of blade bearing are obtained by the modified Unified Jacobian-Torsor model with considering the geometric tolerances and dimensional tolerances; the thermo mechanical coupling deformations of blade bearing of CPP caused by centrifugal force, bolt preload and temperature are obtained by finite element analysis. Considering the friction and wear during service life, the prediction model of service life of blade bearing with considering the time-variant deviation is constructed subject to the constrains of sealing performance and flexibility of adjusting the pitch. Through this method, the influences of temperature, speed and preload of bolt on the assembly deviation are analyzed. It is also found that the thermo mechanical coupling load has great influence on the assembly deviation and service life. Especially, if the temperature and rotation speed increases, the service life decreases largely.

Based on the work of this paper, in the future, the assembly parameters would be optimized with considering the time-variant deviation, and the optimized results would be more reliable. It will also be useful for engineers to design tolerances for parts with considering the real working condition.

Acknowledgments

This research was founded by Major Project of Chinese National Programs for Fundamental Research and Development (Grant No.2014CB046705), all support is gratefully acknowledged.

References

[1] Picheral, L., K. Hadj-Hamou, and J. Bigeon, *Robust optimization based on the Propagation of Variance method for analytic design models.* International Journal of Production Research, 2014. 52(24): p. 7324-7338.

[2] Laperriere, L. and T. Kabore, *Monte Carlo simulation of tolerance synthesis equations.* International Journal of Production Research, 2001. 39(11): p. 2395-2406.

[3] Hong, Y.S. and T.C. Chang, *Tolerancing algebra: a building block for handling tolerance interactions in design and manufacturing - Part 1: Concept and representation.* International Journal of Production Research, 2002. **40**(18): p. 4633-4649.

[4] Pierre, L., D. Teissandier, and J.P. Nadeau, *Integration of thermomechanical strains into tolerancing analysis.* International Journal on Interactive Design and Manufacturing (IJIDeM), 2009. **3**(4): p. 247-263.

[5] Pierre, L., D. Teissandier, and J.P. Nadeau, Variational tolerancing analysis taking thermomechanical strains into account: Application to a high pressure turbine. Mechanism and Machine Theory, 2014. **74**: p. 82-101.

[6] Benichou, S. and B. Anselmetti, *Thermal dilatation in functional tolerancing.* Mechanism and Machine Theory, 2011. **46**(11): p. 1575-1587.

[7] Grandjean, J., Y. Ledoux, and S. Samper, *On the role of form defects in assemblies subject to local deformations and mechanical loads.* The International Journal of Advanced Manufacturing Technology, 2013. **65**(9-12): p. 1769-1778.

[8] Mazur, M., M. Leary, and A. Subic, Computer Aided Tolerancing (CAT) platform for the design of assemblies under external and internal forces. Computer-Aided Design, 2011. **43**(6): p. 707-719.

[9] Mazur, M., M. Leary, and A. Subic, Application of Polynomial Chaos Expansion to Tolerance Analysis and Synthesis in Compliant Assemblies Subject to Loading. Journal of Mechanical Design, 2015. 137(3).

[10] Godjevac, M., et al., *Prediction of fretting motion in a controllable pitch propeller during service.* Proceedings of the Institution of Mechanical Engineers, Part M: Journal of Engineering for the Maritime Environment, 2009. **223**(4): p. 541-560.

[11] Roy, U. and B. Li, *Representation and interpretation of geometric tolerances for polyhedral objects. II.: Size, orientation and position tolerances.* Computer-Aided Design, 1999. **31**(4): p. 273-285.

[12] Chen, H., et al., *A Modified Method of the Unified Jacobian-Torsor Model for Tolerance Analysis and Allocation.* International Journal of Precision Engineering and Manufacturing, 2015. **16**(8): p. 1789-1800.

[13] Ghie, W., L. Laperrière, and A. Desrochers, *Statistical tolerance analysis using the unified Jacobian–Torsor model.* International Journal of Production Research, 2010. **48**(15): p. 4609-4630.

[14] Godjevac, M., *Wear and friction in a controllable pitch propeller.* 2010: TU Delft, Delft University of Technology.

IOP Conf. Series: Materials Science and Engineering **212** (2017) 012022 doi:10.1088/1757-899X/212/1/012022

Due Date Assignment in a Dynamic Job Shop with the Orthogonal Kernel Least Squares Algorithm

D.H. Yang, L. Hu, Y. Qian

School of Business Administration, Shenzhen Institute of Information Technology, Shenzhen, China

yangdh@sziit.edu.cn; hulingedu@126.com; 4091172@qq.com

Abstract. Meeting due dates is a key goal in the manufacturing industries. This paper proposes a method for due date assignment (DDA) by using the Orthogonal Kernel Least Squares Algorithm (OKLSA). A simulation model is built to imitate the production process of a highly dynamic job shop. Several factors describing job characteristics and system state are extracted as attributes to predict job flow-times. A number of experiments under conditions of varying dispatching rules and 90% shop utilization level have been carried out to evaluate the effectiveness of OKLSA applied for DDA. The prediction performance of OKLSA is compared with those of five conventional DDA models and back-propagation neural network (BPNN). The experimental results indicate that OKLSA is statistically superior to other DDA models in terms of mean absolute lateness and root mean squares lateness in most cases. The only exception occurs when the shortest processing time rule is used for dispatching jobs, the difference between OKLSA and BPNN is not statistically significant.

Keywords: Due Date Assignment; Job Shop; Orthogonal Kernel Least Squares Algorithm

1. Introduction

In order to remain competitive in today's global markets, modern manufacturing systems often work in dynamic environments. Aside from providing high quality products, meeting due dates is another crucial requirement of production control. The "due date" is the specified date when production of a particular order is expected to be complete [1]. No matter imposed externally by the customers or set internally by the manufacturer, the due date related performance has a great impact upon the economic benefit of the production [2]. A good due date assignment (DDA) rule can perform timely delivery and reduce inventory costs, thereby enhancing the customer satisfaction and competitive advantage [3]. A typical DDA procedure is to estimate the flow-time of an arriving job and then adding that to its arrival time. It is very simple to estimate the job flow-time and set the due date in a purely deterministic system. However, estimating the accurate flow-time is not an easily-solved problem in a dynamic environment, due to the stochastic nature of arriving jobs and the uncertainty of processing sequences caused by some dispatching rules.

Owing to the critical importance of due dates, the DDA problem in a dynamic, multi-machine manufacturing system has been studied for a long time. During the first years, studies were focused on simple, regression based approaches. Some conventional rules, e.g. TWK, NOP, TWK+NOP, JIQ, WIQ, WEEKS, JIS and RMR, were proposed by different researchers and a thorough comparison was conducted in [2]. These rules employ different kinds of job characteristics and shop state information to predict job completion times. The comparison results demonstrate that information related to the machine congestion along a job's routing is more useful than general shop conditions.

In general, the aforementioned conventional DDA rules rely on linear or low-order nonlinear regression methods. They lack the ability to reveal the complex nonlinear relationship between actual job flow-times and relevant information concerning job characteristics and system state. To overcome the shortcoming of these rules, a number of researchers subsequently applied artificial intelligence approaches to assign due dates in job shops or wafer fabrication. Neural Networks (NNs) were first employed to specify the internally-set due dates in dynamic job shops [4]. After that, NNs continued to show effectiveness and robustness when they were combined with other techniques in varying production environments [5, 6]. Except for NNs, other artificial intelligence approaches were also incorporated to improve the DDA performance, such as Self-Organizing Map and fuzzy rule [7].

In this paper, we propose a DDA method by using the Orthogonal Kernel Least Squares Algorithm (OKLSA) [8, 9]. OKLSA aims at finding an approximate solution for the kernel least squares problem and obtains a regression model with a sparse structure. A discrete-event simulation model is first built to imitate the production process of a highly dynamic job shop. A series of simulation runs are then conducted with varying dispatching rules (FIFO, SPT, EDD) to collect data set for training, validating and testing different DDA models. Some numerical experiments are carried out to evaluate the DDA performance of OKLSA by comparing its prediction result with five conventional rules and Back-Propagation Neural Network (BPNN) in terms of Mean Absolute Lateness (MAL) and Root Mean Squares Lateness (RMSL) criteria. The experimental results indicate that OKLSA statistically outperforms other DDA models in most cases. The only exception occurs when SPT rule is used for dispatching jobs, the difference between OKLSA and BPNN is not statistically significant.

The remainder of the paper is organized as follows. Section 2 reviews the basic theory of OKLSA. Section 3 describes the OKLSA-based methodology for due date assignment. Furthermore, section 4 introduces the experimental design and data generation. Section 5 shows the experimental results and gives a brief analysis. The conclusions and suggestions for future work are presented in Section 6.

2. Orthogonal Kernel Least Squares Algorithm[8, 9]

2.1. Kernel Least Squares Algorithm

Assume to be given a training data set $\{(\mathbf{x}_i, y_i)\}_{i=1}^n$ with input vectors $\mathbf{x}_i \in \Re^N$ and associated target outputs $y_i \in \Re$. Kernel Least Squares Algorithm (KLSA) consists of finding a function:

$$f(\mathbf{x}) = \mathbf{w}^T \varphi(\mathbf{x}) + b, \tag{1}$$

where the mapping $\varphi(\cdot)$ is implemented via a kernel that satisfies the Mercer's condition such that $K(\mathbf{x}_i, \mathbf{x}_j) = \varphi(\mathbf{x}_i)^T \varphi(\mathbf{x}_j)$. The output y_i can be expressed as:

$$y_i = f(\mathbf{x}) + \xi_i = \mathbf{w}^T \varphi(\mathbf{x}) + b + \xi_i, \tag{2}$$

where ξ_i is the training error. In accordance with the Representer Theorem [10], the coefficient vector $\mathbf{w}$ of the optimal solution of (1) must lie in the span of all training data points in the feature space, in the form of

$$\mathbf{w} = \sum_{i=1}^n \alpha_i \varphi(\mathbf{x}_i) \tag{3}$$

where $\alpha_i\ (i = 1,...,n)$ are the coefficients. Substituting (3) into (2) and omitting the threshold yields:

$$y_i = f(\mathbf{x}) + \xi_i = \sum_{i=1}^n \alpha_i K(\mathbf{x}_i, \mathbf{x}) + \xi_i . \tag{4}$$

Expression (4) can be reformulated with the concise form

$$\mathbf{y} = \mathbf{K}\boldsymbol{\alpha} + \boldsymbol{\Xi}, \tag{5}$$

where $\mathbf{y}=[y_1,...,y_n]^T$, $\mathbf{K}=[\mathbf{k}_1,...,\mathbf{k}_n]=\begin{bmatrix} K(\mathbf{x}_1,\mathbf{x}_1) & \cdots & K(\mathbf{x}_1,\mathbf{x}_n) \\ \vdots & \ddots & \vdots \\ K(\mathbf{x}_n,\mathbf{x}_1) & \cdots & K(\mathbf{x}_n,\mathbf{x}_n) \end{bmatrix}$, $\boldsymbol{\alpha}=[\alpha_1,...,\alpha_n]^T$, and $\boldsymbol{\Xi}=[\xi_1,...,\xi_n]^T$.

The regression model can be obtained by minimizing the sum of squared residual errors as:

$$\min \ \|\mathbf{K}\boldsymbol{\alpha}-\mathbf{y}\|^2. \tag{6}$$

2.2. Orthogonal Kernel Least Squares Algorithm

The computational complexity and memory requirement for computing the solution of (6) is prohibitive for very large datasets. In addition, the "dense" structure of the yielded model is easy to cause the overfitting problem. An alternative approach to overcome these difficulties would be to enforce regularization via “sparsity”, with which the coefficient vector $\mathbf{w}$ in (3) can be approximated by

$$\mathbf{w} \approx \sum_{j=1}^{m} \alpha_j \varphi(\mathbf{x}_j^*) \tag{7}$$

where $m\ (<n)$ is a positive integer and $\mathbf{x}_j^* \in \{\mathbf{x}_i, i=1,...,n\}$. Problem (6) hence yields an approximation problem as:

$$\min \ \|\mathbf{K}^m\boldsymbol{\alpha}^m-\mathbf{y}\|^2, \tag{8}$$

where $\boldsymbol{\alpha}^m=[\alpha_1,...,\alpha_m]^T$, $\mathbf{y}=[y_1,...,y_n]^T$, and $\mathbf{K}^m=\begin{bmatrix} K(\mathbf{x}_1,\mathbf{x}_1^*) & \cdots & K(\mathbf{x}_1,\mathbf{x}_m^*) \\ \vdots & \ddots & \vdots \\ K(\mathbf{x}_n,\mathbf{x}_1^*) & \cdots & K(\mathbf{x}_n,\mathbf{x}_m^*) \end{bmatrix}$ being a rectangular matrix formed by choosing a subset of m columns from the original kernel matrix $\mathbf{K}$. The final function is then expressed as $f(\mathbf{x})=\sum_{j=1}^{m}\alpha_j K(\mathbf{x}_j^*,\mathbf{x})$. When m is a small number, the obtained regression model has a sparse structure. Since it is computationally demanding to obtain the optimal solution of (8), research efforts have been concentrated on developing methods to obtain sub-optimal solutions.

To get a sub-optimal solution of (8), OKLSA attempts to select a subset of the most significant terms for the regression model in a sequential forward manner. It calculates the individual contribution of each kernel term $\varphi(\mathbf{x}_i)$ to error minimization by transforming the set of $\mathbf{k}_i$ $(i=1,...,n)$ into a set of orthogonal vectors. Hence, the matrix $\mathbf{K}$ is decomposed into the product of two matrices represented as $\mathbf{K}=\mathbf{UB}$, where $\mathbf{B}\in\Re^{n\times n}$ is an upper-triangular matrix with 1's on the diagonal and $\mathbf{U}\in\Re^{n\times n}$ is an matrix with orthogonal columns such that $\mathbf{U}^T\mathbf{U}=\mathbf{T}$, where $\mathbf{T}$ is diagonal and the diagonal elements are calculated by $t_i=\mathbf{u}_i^T\mathbf{u}_i$. Therefore, the space spanned by the set of orthogonal vectors $\mathbf{u}_i$ is the same as the space that is spanned by the set of $\mathbf{k}_i$. Equation (5) can be rewritten as:

$$\mathbf{y}=(\mathbf{K}\mathbf{U}^{-1})(\mathbf{U}\boldsymbol{\alpha})+\boldsymbol{\Xi}=\mathbf{Tg}+\boldsymbol{\Xi}, \tag{9}$$

When minimizing the total squared errors $\boldsymbol{\Xi}^T\boldsymbol{\Xi}$, its solution is given by

$$\mathbf{g}=\mathbf{T}^{-1}\mathbf{U}^T\mathbf{y}, \tag{10}$$

or $g_i=\mathbf{u}_i^T\mathbf{y}/(\mathbf{u}_i^T\mathbf{u}_i)$ $(i=1,...,n)$. The values of the vectors $\mathbf{g}$ and $\boldsymbol{\alpha}$ satisfy $\mathbf{B}\boldsymbol{\alpha}=\mathbf{g}$. Knowing $\mathbf{B}$ and $\mathbf{g}$, $\boldsymbol{\alpha}$ can be determined through back substitution. Due to the fact that $\mathbf{u}_i$ and $\mathbf{u}_j$ are orthogonal for $i\neq j$, the total squares errors can be represented as:

$$\boldsymbol{\Xi}^T\boldsymbol{\Xi}=\mathbf{y}^T\mathbf{y}-\sum_{i=1}^{n} g_i^2\mathbf{u}_i^T\mathbf{u}_i \tag{11}$$

Equation (11) implies that the error reduction due to $\mathbf{u}_i$ is defined as:

$$[err]_i = g_i^2 \mathbf{u}_i^T \mathbf{u}_i / (\mathbf{y}^T \mathbf{y}). \tag{12}$$

OKLSA selects the kernel terms one by one according to their contributions to residual error minimization. According to (12), it is convenient to select a couple of most significant kernel terms, which leads to the greatest reduction in the residual errors. When applying OKLSA to solving (8) , the selection of a subset of significant kernel terms is equivalent to the selection of a subset of columns from the kernel matrix.

3. OKLSA-based Methodology for Due Date Assignment

The due date of a given job should be equal to the sum of the job's arrival time and flow-time in the system. This can be represented by the equation: $D_i = A_i + F_i$, where D_i represents the due date of job *i*, A_i is its arrival time, and F_i is its flow-time. The typical DDA procedure therefore consists of estimating the flow-time of the arriving job and then adding the estimated flow-time to the job's arrival time accordingly. The methodology for DDA with OKLSA can be described as follows:

(1) A discrete-event simulation model is first tailored according to the system configuration and the dispatching rule used for controlling the job sequence.
(2) For each particular set of information inputs, a simulation is run for a lengthy period to generate a set of data. The whole dataset is segmented into three subsets: the training set, the validation set and the test set.
(3) A DDA model is subsequently constructed with OKLSA based on the training data.
(4) The data in the validation set is used to determine the optimal parameters in OKLSA and the effectiveness of the obtained DDA model is evaluated by the data in the test set.

4. Experimental Design

To study the prediction performance of various DDA models, we build a discrete-event simulation model with C++ language to imitate the production process of a highly dynamic job shop. The shop has nine machines and the size of each machine buffer is assumed to be unlimited. The job inter-arrival times are drawn from an exponential distribution. The desired shop utilization rate is yielded by adjusting the job arrival rate. The characteristics of each job are generated on arrival. Jobs entering the shop are randomly routed with no consecutive operations on the same machine permitted. The number of operations per job is drawn from a uniform distribution in the range from 1 to 9. The operation times are drawn from an exponential distribution with a mean of 1. The set-up times are included in the operation times and the transportation times are negligible. It is assumed that there are no mechanical failures or required maintenance on all the machines during the simulation horizon.

To evaluate the relative effectiveness and robustness of DDA models under varying dispatching rules, three different dispatching rules (FIFO, SPT and EDD) are used for controlling the job sequence. The shop utilization level is kept at 90%, which indicates that the work load in the shop is relatively heavy. For a particular system situation, a simulation is run to generate sufficient data examples for training, validation and test. Each data example is an "attributes – target value" pair. The "attributes" are those variables describing job characteristics and system state information. The "target value" refers to the job flow-time. Seven independent attributes are selected as [4] and listed in Table 1.

Once the simulation begins, the first 500 jobs are treated as a warm-up period. When the system reaches the steady state, data recordation begins. At the instant job *i* arrives at the system, the aforementioned seven attributes are recorded and form a vector $\mathbf{x}_i$. When job *i* is completed and exits the system, its flow-time through the system is recorded as a scalar y_i. Hence, each data example is notated in an "attributes – target value" pair represented as $(\mathbf{x}_i, y_i)$. To reduce the possibility of serial correlation and guaranteeing statistical independence, one data example is selected from every twenty jobs. A total of 13,000 data examples are collected. All data examples are randomly permuted and independently divided into three sets. The first set includes 1,000 data examples as training data, which are used to estimate the regression parameters of the five conventional rules as well as to train BPNN and OKLSA. The second set involving 2,000 data examples is the validation set to determine the optimal parameters used in BPNN and OKLSA. The remaining 10,000 data examples comprise the third set used for evaluating the prediction performance of various DDA models. Moreover, to avoid the numerical problems in OKLSA and BPNN, each attribute is linearly scaled to the range [-1,+1].

Table 1. The selected attributes describing job characteristics and system state information.

Index	Attributes description
1	Number of operations required by job *i*
2	Total processing time for job *i*
3	Number of operations that must be done by the machine on job *i*'s path to complete all jobs presently in the shop
4	Sum of jobs presently in queues on job *i*'s routing
5	Maximum processing time job *i* requires on any machine
6	Number of operations required to empty the shop of its current workload
7	Total processing time of jobs in queues on job *i*'s routing

5. Experimental Results and Analysis

In the experiments, the prediction performance of OKLSA is compared with those of the aforementioned five conventional rules (i.e. TWK, NOP, TWK+NOP, JIQ and WIQ) and BPNN. Mean Absolute Lateness (MAL) and Root Mean Squares Lateness (RMSL) are chosen as the criteria [11, 12]. MAL and RMSL are defined respectively as: $MAL=\sum_{i=1}^{n}|L_i|/n$, $RMSL=\sqrt{\sum_{i=1}^{n}L_i^2/n}$, where $L_i = AF_i - PF_i$, n is the total number of jobs, L_i is the lateness of job *i*, AF_i refers to the actual flow-time of job *i*, and PF_i denotes the predicted flow-time of job *i*. The coefficients of the five conventional rules are calculated by minimizing the variance between predicted flow-times and actual flow-times among the training data. BPNN employs a common three-layer neural network with the 7-10-1 structure and is trained by the back-propagation algorithm. Sigmoid transfer function is used in the hidden-layer, and linear function is used in the output-layer. The RBF kernel $K(\mathbf{x}_i,\mathbf{x}_j)=\exp(-\gamma\|\mathbf{x}_i-\mathbf{x}_j\|^2)$ is adopted in OKLSA, where γ is selected in the range $\gamma\in\{2^{-7},2^{-6},\ldots 2^{1},2^{2}\}$. The number of kernel terms used in OKLSA is selected in the range $m\in\{4,6,8,10,12,14,16,20,22,24,26,28,30,32,35\}$. The optimal set of parameters (of OKLSA or BPNN) is determined when the prediction error on the validation set is minimized. This set of parameters is then used to train a DDA model for evaluating the prediction error on the test set. The prediction results of various DDA models are shown in Tables 2(a) and 2(b). In the tables that list MAL or RMSL, each value represents the average value of 10,000 test data examples. The best prediction values are written in the bold and italic form.

Table 2(a). Prediction results in terms of MAL.

	FIFO	*SPT*	*EDD*
TWK	22.95	12.68	14.21
NOP	19.02	14.52	17.57
TWK+NOP	18.96	12.75	13.92
JIQ	9.40	12.65	9.93
WIQ	8.37	12.72	10.03
BPNN	8.33	***11.15***	7.70
OKLSA	***8.05***	11.22	***7.50***

Table 2(b). Prediction results in terms of RMSL.

	FIFO	**SPT**	**EDD**
TWK	31.69	32.97	18.98
NOP	25.73	36.34	22.80
TWK+NOP	25.62	32.32	18.14
JIQ	13.76	32.96	12.87
WIQ	12.80	32.97	13.03
BPNN	12.48	30.69	10.44
OKLSA	***12.07***	***30.42***	***9.95***

Just like the prior research [4], the statistical test (the paired *t*-test) is carried out to check whether OKLSA is significantly superior to the other methods. The 10,000 test data examples are grouped into ten batches and paired *t*-tests are then conducted using the batch means to test the hypotheses. Since the number of test data examples is large, the MAL or RMSL of each batch are regarded as being approximately uncorrelated and normally distributed according to the central limit theorem. This condition thereby meets the assumptions of the paired *t*-test. One-tailed hypotheses are formulated that the MAL or RMSL for each of the conventional rules or BPNN do not exceed that of OKLSA. The hypotheses are hoped to be rejected. In particular: $H_0 : \mu_D \leq 0$; $H_1 : \mu_D > 0$,where D denotes the MAL (or RMSL) difference between the conventional rules (or BPNN) and OKLSA. For all hypotheses tests, the level of significance is set at $\alpha = 0.05$. The Bonferroni's inequality is used to control the experiment-wise error rate of the six multiple comparisons. Therefore the *p*-value must be less than $0.05/6 = 0.0083$ for comparisons to be statistically significant. Results of the statistical tests are listed in the form of *t*-values and *p*-values for the MAL or RMSL criterion in Tables 3(a) and 3(b).

Table 3(a). Statistical test *t*-values and *p*-values (MAL).

		FIFO		SPT		EDD	
		t	*p*	*t*	*p*	*t*	*p*
OKLSA versus	TWK	29.0562	0.0001	13.5121	0.0001	16.7889	0.0001
	NOP	31.8841	0.0001	15.6180	0.0001	32.6923	0.0001
	TWK+NOP	30.6449	0.0001	17.8034	0.0001	16.1800	0.0001
	JIQ	23.5956	0.0001	13.1993	0.0001	16.8179	0.0001
	WIQ	6.0939	0.0001	14.3742	0.0001	16.4340	0.0001
	BPNN	9.3715	0.0001	-0.7568	0.7657	6.6434	0.0001

Table 3(b). Statistical test *t*-values and *p*-values (RMSL).

		FIFO		SPT		EDD	
		t	*p*	*t*	*p*	*t*	*p*
OKLSA versus	TWK	23.6865	0.0001	4.4193	0.0008	14.4895	0.0001
	NOP	19.7061	0.0001	8.2708	0.0001	26.3573	0.0001
	TWK+NOP	19.0961	0.0001	4.1633	0.0012	13.8695	0.0001
	JIQ	20.2452	0.0001	4.4397	0.0008	12.8460	0.0001
	WIQ	6.8000	0.0001	4.3797	0.0008	12.3404	0.0001
	BPNN	7.1634	0.0001	0.9039	0.1948	6.7368	0.0001

The hypotheses test results show that the calculated *p*-values are less than 0.0083 in most cases. It is clearly indicated that the null hypothesis is rejected for most hypotheses at an alpha-level of 0.05 and OKLSA statistically outperforms the conventional rules and BPNN in most cases. OKLSA is able to predict job flow-times with higher accuracy than the five conventional rules in all cases. Due to the employment of kennel function, OKLSA is able to construct nonlinear models to reveal the complicated relationship that exists between the job flow-times and the influential factors. All the conventional DDA rules use linear regression methods to construct simple DDA models. Although they are easy to implement and comprehend, their performance is relatively poor mainly because such linear models are easy to cause the "underfitting" problem. Hence, it is reasonable that OKLSA is significantly superior to the conventional rules.

OKLSA also substantially surpasses BPNN, with the only exception when SPT being used for dispatching jobs within the system. The advantage of OKLSA over BPNN lies in two points: (1) the use of kernel function enables OKLSA to learn the training data in the high-dimension feature space; and (2) OKLSA is able to construct models with a sparse structure, which makes OKLSA easily generalize the learned knowledge to the newly arrived data. In the case that SPT is employed, OKLSA shows a similar prediction accuracy to BPNN on the RMSL criterion but a little worse than BPNN on the MAL criterion, although the difference is not statistically significant. When the SPT rule is applied to sequencing job operations, some jobs with vary large lateness are likely to occur. When there are a large number of such very-late jobs in the data set, the prediction performance of OKLSA may be heavily influenced by them.

6. Conclusions and Future Research

This paper constructs a sparse nonlinear regression model to assign due date in a dynamic job shop by using the Orthogonal Kernel Least Square Algorithm. A discrete-event simulation model has been built and some simulation experiments have been carried out to evaluate the performance of OKLSA by comparing its prediction accuracy with other DDA models. The experimental results indicate that OKLSA statistically outperforms all the conventional DDA rules and BPNN with regard to both MAL and RMSL criteria in most cases. The only exception occurs when the SPT rule is used for dispatching jobs within the system. Influenced by very-late jobs, OKLSA shows similar performance to BPNN on the RMSL criterion but is slightly poorer than BPNN on the MAL criterion, although the difference is not statistically significant. Due to the desirable prediction performance achieved in most cases, OKLSA is therefore proven to be a very promising methodology for DDA in the dynamic job shops.

In this study, OKLSA has not shown the best performance when the SPT rule is used in the job shop. It is thus important to develop an effective strategy or technique in the future to enhance OKLSA's performance in the SPT case. Moreover, OKLSA suffers from expensive computational costs when the dataset is very large. One promising area for further research would be to derive more efficient training algorithms and apply them to solve DDA problems.

Acknowledgments
In this paper, the research was sponsored by the "The 1st-Class Excellent High Vocational College Project of Guangdong Province", "The 6th Teaching Research Project of Shenzhen Institute of Information Technology (Project No. 2016jgyb04) " and "The 13th Five-year Plan Education Project in Shenzhen (Project No. ybzz16009)".

References

[1] Eilon, S. and I.G. Chowdhury, Due dates in job shop scheduling. The International Journal of Production Research, 1976. 14(2): pp. 223-237.
[2] Ragatz, G.L. and V.A. Mabert, A simulation analysis of due date assignment rules. Journal of Operations Management, 1984. 5(1): pp. 27-39.
[3] Sha, D.Y. and C.H. Liu, Using Data Mining for Due Date Assignment in a Dynamic Job Shop Environment. International Journal of Advanced Manufacturing Technology, 2005. 25(11-12): pp. 1164-1174.
[4] Philipoom, P.R., L.P. Rees and L. Wiegmann, Using neural networks to determine internally-set due-date assignments for shop scheduling. Decision Sciences, 1994. 25(5): pp. 825-851.
[5] Hsu, S.Y. and D.Y. Sha, Due date assignment using artificial neural networks under different shop floor control strategies. International Journal of Production Research, 2004. 42(9): pp. 1727-1745.
[6] Patil, R.J., Using ensemble and metaheuristics learning principles with artificial neural networks to improve due date prediction performance. International Journal of Production Research, 2007. 46(21): pp. 6009-6027.
[7] Chang, P. and T.W. Liao, Combining SOM and fuzzy rule base for flow time prediction in semiconductor manufacturing factory. Applied Soft Computing, 2006. 6(2): pp. 198-206.
[8] Billings, S.A. and K.L. Lee, Nonlinear Fisher discriminant analysis using a minimum squared error cost function and the orthogonal least squares algorithm. Neural Networks, 2002. 15(2): pp. 263-270.
[9] Nair, P.B., A. Choudhury and A.J. Keane, Some greedy learning algorithms for sparse regression and classification with mercer kernels. The Journal of Machine Learning Research, 2002. 3: pp. 781-801.

[10] Kimeldorf, G. and G. Wahba, A Correspondence Between Bayesian Estimation on Stochastic Processes and Smoothing by Splines. Annals of Mathematical Statistics, 1970. 41(2): pp. 495-502.
[11] Vig, M.M. and K.J. Dooley, Dynamic rules for due-date assignment. The International Journal of Production Research, 1991. 29(7): pp. 1361-1377.
[12] Sabuncuoglu, I. and A. Comlekci, Operation-based flowtime estimation in a dynamic job shop. Omega, 2002. 30(6): pp. 423-442.

Comparison and Evolution of Energy Consumption in Moroccan Agro-food Industries

Meryem EL BADAOUI[1], Abdellatif TOUZANI[2]

Applied Thermodynamics and Solid Combustibles (ATSC), Mohammedia School of Engineers (EMI), Avenue Ibn Sina BP: 765, Agdal- Rabat Morocco

[1]meryemelbadaoui@research.emi.ac.ma ;[2]atouzani@emi.ac.ma

Abstract. The aim of this article is to establish a comparison between the Moroccan energy consumption and the BREF the reference document on best available techniques in the food industries, then an evolution of this consumption by 2030 in order to better understand it and to define strategies to reduce energy bill. According to a survey conducted among 5000 Moroccan companies, we were able to compare the energy consumption of the agro-food industries including sugar industry, dairy industry, cereal industry; fatty substances industry and fishing industrywith that of the BREF. Also an evolution of Moroccan consumption was established by 2030 using the linear regression method, and then calculated a non-negligible average annual growth rate (AAGR).The results show that the Moroccan energy consumption is adequate to that of the BREF, and an energy consumption constantly increasing by registering a non-negligible AAGR.

1. Introduction

The sector of agro-alimentary industries (IAA) occupies the second place in the Moroccan industry after the chemical and Para chemical industries, it accounts for nearly 30% of total industrial production [1].

In 2013, this large sector has 2062 companies, and employs 140,975 people. Thus an energy consumption recording a value of 378.182 toe of which 130.499 toe electrical energy and 247.683 toe thermal energy [2]. The sugar industry, the dairy industry, the cereal industry, the fats and oils industry and the fishing industry count among the most energy-consuming food processing industries by counting approximately 55% of total energy consumption in the agro-food sector [3]. In order to reduce the consumption of the sector, Morocco aims by 2030 to minimize it by implementing various strategies to cite Green Morocco Plan.

2. Methodology

Our article is based on the results of a survey conducted among 5000 Moroccan companies including the agro-food industries focusing on their energy consumption and comparing it with that of the BREF so as to better know the energy consumption of the sector.

In order to predict the energy consumption, the following linear regression method was used for the agro-food industries: the sugar industry, dairy industry, cereal industry, fatty industry and fishing industry.

- Dairy Industry
  - Electric energy consumption: y = 0,27x - 536,38
  - Thermal energy consumption: y = 0,452x - 897, 45

Where "x" is the year of energy consumption and "y" is the energy consumption including electric and thermal.

- Cereal Industry
  - Electric energy consumption: y = -0,83x + 1706, 8
- Fat industry
  - Olive oil industry
    - Electric energy consumption – Olive oil: y = 0,3813x - 764, 21
    - Electric energy consumption – Table olive: y = 0,6963x - 1398,7
  - Oilseed industry
    - Electric energy consumption: y = 0,079x – 158,45
    - Thermal energy consumption: y = 0,203x - 407,18
- Fishing industry
  - Electric energy consumption: y = 0,007x - 11,59
  - Thermal energy consumption: y = 0,029x - 48,38

3. Results and Discussion

This section is interested in presenting the results, as well as their discussions. First, we present below the comparison of the energy consumption, including the electrical and thermal consumption with that of the BREF, and then the consumption evolution is predicted using the linear regression method cited above.

Also, the AAGR is calculated using the following formula; $\mathbf{AAGR} = \left(\sqrt[n]{\frac{\mathbf{EV}}{\mathbf{FV}}} - \mathbf{1}\right) * \mathbf{100}$

With "EV": End value, refers to the value of the later year.

"FV": First value refers to the value of the earlier year.

"n" : Number of periods between the earlier and the later periods.

This indicator measure the average evolution in energy consumption from 2013 to 2030

3.1. Sugar industry

The sugar industry is one of the major consumers of energy to quote 44% of the total energy consumption. Table 1 and Table 2 show the total energy consumed in the sugar industry according to the BREF and the Moroccan one, while Figure 1 represents the evolution of the Moroccan industry's energy consumption by 2030.

Table1. Total energy consumed in the sugar industry according to the BREF

Total energy consumed (KWh)			
Specific value per ton of beet		Specific value per ton of sugar produced	
Average	Range	Average	Range
307	232-367	1987	1554-2379

Table 2. Energy consumed per ton of sugar produced in Moroccan industries

Products	Electricity (KWh/t)	Fuel (KWh/t)
Sugar industry	39	2242

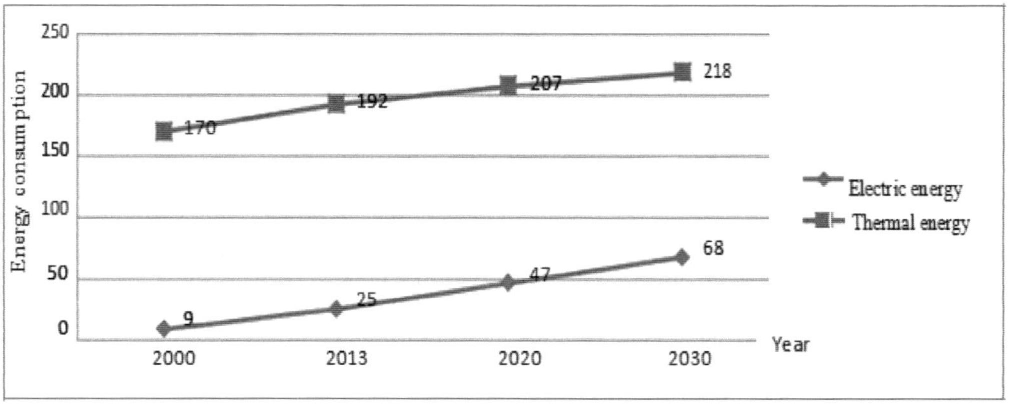

Figure 1. Evolution of Moroccan energy consumption in 1000 Toe

According to the results of Tables 1 and 2, the total energy consumed per ton of sugar produced in Moroccan industries amounts to 2281 kWh, which is in line with the range delivered by the BREF with respect to figure 1 whose consumption of electricity, as well as fuel by 2030, has an average annual growth rate of around 6.97% for the first consumption. While the second one mark a humble AAGR of 0.83%. The latter Rate is the result of the integration of other energy alternatives, namely the valued bagasse thus re-use it in order to minimize the use of fossil energy as well as a reduction of the energy bill of this sector.

3.2. Dairy industry

In order to extract dairy products, energy consumption is essential, recording a consumption of around 6% of the total energy consumption of the agro-food sector. 80% of this energy consumption is in the form of thermal energy. The rest is in the form of electrical energy. Table 3 shows the Moroccan energy consumption and that of the BREF of dairy products. Regarding figure 2, it shows us an evolution of the energy consumption by 2030.

Table3. Energy Consumption of Dairy Products

Products	Electricity (KWh/t) Moroccan consumption	Range (BREF)	Fuel (KWh/t) Moroccan consumption	Range (BREF)
Milk and yogurt marketed	46	42-694	178	50-417
Cheese	417	22-805	410	42-1278
Milk and milk powder	66	17-917	233	333-5555

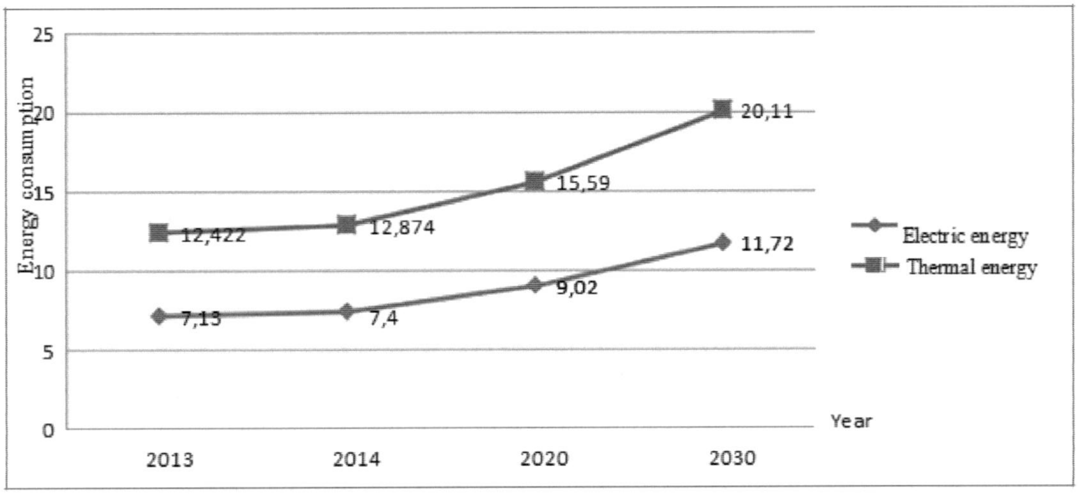

Figure2. Evolution of Moroccan energy consumption in 1000 Toe

Table 3 above shows that the Moroccan energy consumption respects the data of the BREF except the consumption of thermal energy of milk and milk powder, which is a little less than the range. Concerning figure 2 Electricity and fuel consumption are constantly increasing, with an average annual growth rate of around 2.96% for the first consumption, while the second marks an AAGR of 2.87%. This is due to the recent policy adopted by the sector to minimize their energy consumption [4].

3.3. *Cereal industry*

The cereal sector is one of the main sectors of agricultural production in Morocco, recording a consumption of 27% of the total electricity consumption of the agro-food sector. Table 4 shows energy consumption of the pasta according to the BREF, while figure 3 shows the evolution of the electrical consumption of the sector.

Table 4. Energy consumption of pasta according to BREF

Energy consumption	Range (KWh/t)
Electric	140-220
Thermal	0,417-0,527

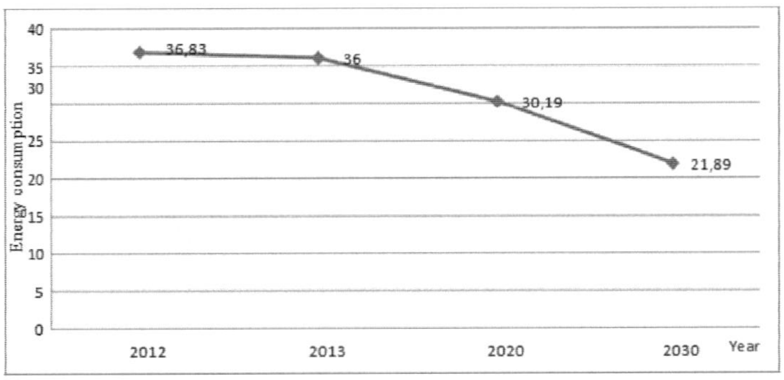

Figure 3. Electricity consumption for Moroccan cereal industry in 1000 Toe

Concerning figure 3, electricity consumption for the Moroccan cereal industry is down by -2.85%. This can be explained by the decrease in production of the three main cereals [5, 6] and the major dependence of this sector on rainfall.

3.4. Fat industry

Considering the importance of oils (in terms of food, energy or industry), the fatty substances industry accounts for a significant percentage in the agro-food transformation by 10.40%, but its energy consumption remains minimal, namely 3% of the total energy consumption of the agro-food sector. Table 5 shows the energy consumption of Moroccan fatty substances. As for figure 4, 5 and 6, they show the increasing evolution of energy consumption in the sector.

Table 5. Energy consumption of the fatty substances industry in Morocco

	Electricity (KWh/t)	Fuel (KWh/t)
Olive oil	209	
Soybean oil, sunflower	116	295

- **Olive oil**

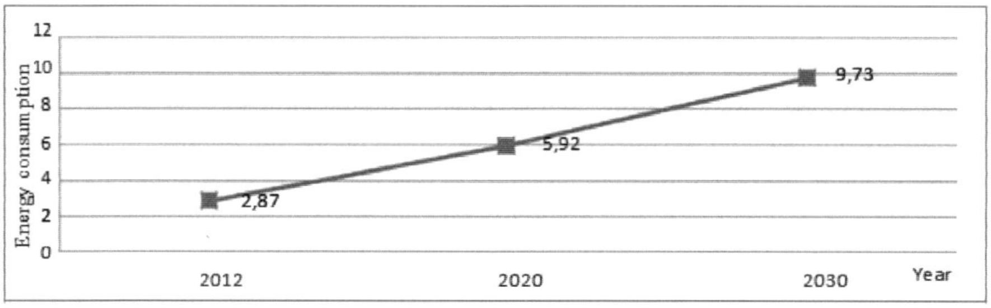

Figure 4. Evolution electricity consumption of Moroccan olive oil in 1000 Toe

- **Table olive**

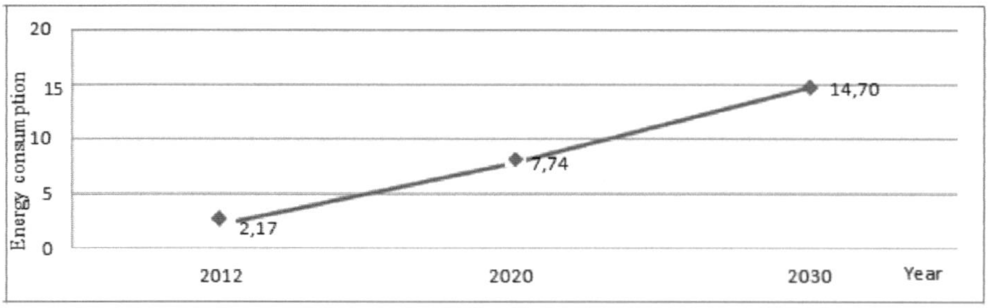

Figure 5. Evolution of electricity consumption of Moroccan table olive in 1000 Toe

➢ **Oilseed industry**

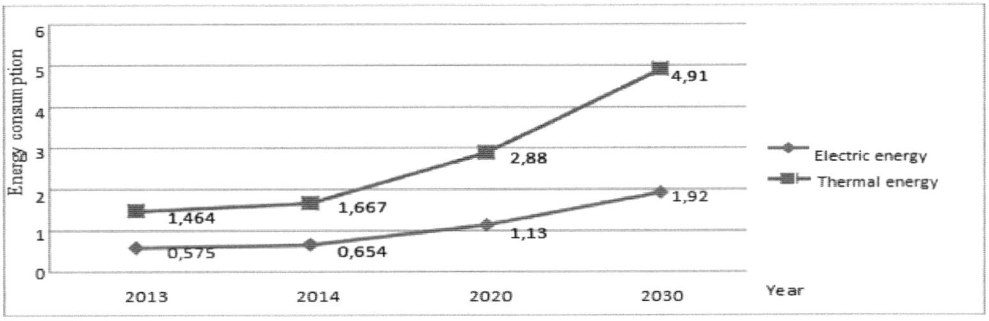

Figure 6. Evolution of energy consumption of Moroccan oilseed industry in 1000 Toe

Table 5 the Moroccan consumption highlights the energy consumed by fatty industry and shows usthat the Moroccan consumption of thermal energy respects the range delivered by the BREF: 155-390 KWh / t while that of Morocco is 294.59 KWh / t. According to figure 4,it shows the evolution of electricity consumption from 2012 to 2030 in olive oil. This consumption continues to increase, registering an average annual growth rate in the order of 7,018%, figure 5 is about the electricity consumption of table olives, it shows an average annual growth rate of about 9.87% and finally figure6 shows us the increase in electricity and fuel consumption for the oilseed industry, with an averageannual growth rate of around 7.35% for first-time consumption, while the second marks a AAGR of 7.64%. These rates obtained from the Moroccan strategy "Green Morocco Plan" concerning the olive oil industry, aim to reach an olive oil production of 330,000 tons by 2020 and a table olive of 320,000 tons, the government as well signed a contract with the inter-professional of oilseed crops FLOEA so as to be able to increase the production of edible oils to reach 95,000 tons in 2020 [7].

3.5. Fishingindustry

According to the National Fisheries Office, Morocco ranks first as a fish producer in Africa, as well as the production and export of sardines at the international level. This sector consumes 10% of the total agribusiness. According to the BREF, the energy consumption during the canning is between 150 and 190 KWh/t while the Moroccan consumption is of 189,84KWh /t the order of electric energy. Figure 7 shows the evolution of Moroccan energy consumption.

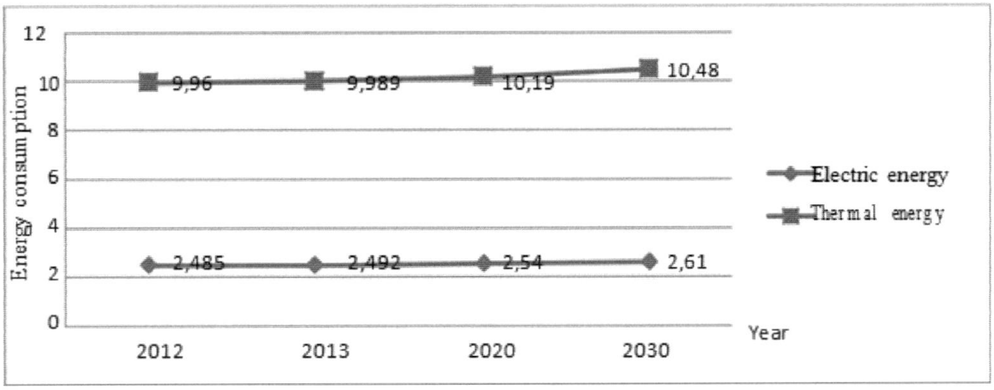

Figure 7. Evolution of energy consumption of Moroccan fishing sector in 1000 toe

Figure 7 shows a slight increase in electricity and fuel consumption in the fishing sector, with amodest annual average growth rate of around 0.27% for the first consumption while the second marks an AAGR of 0.283%. These rates are due to the low Moroccan consumption of sea products, to bearing in mind 11 kg / person / year, compared with 18.9 kg / person / year worldwide [8].

4. Conclusion

The aim of this article is comparing and studying the evolution of Moroccan energy consumption, using the linear regression method. The results obtained from the comparison enabled us to deduce the following conclusions:

- The Moroccan sugar industry has a total energy consumption of around 2281Kwh / t which is respectable by the margin delivered by the BREF[**9**] : [1554-2379]
- Moroccan dairy products also recorded adequate energy consumption to that of the BREF to quote milk and yogurt marketed which their electrical consumption is around of 46 KWh/t and the thermal consumption is 178 KWh/t which are adequate to the margin of the BREF to know [42-694] for electrical consumption then [50-417] for thermal consumption.

On the other hand, they show us that the agro-food industry, represented by its sub-sectors will constantly increase its energy consumption from 378,182 toe in 2013 to 690,000 toe in 2030. The energy bill will also significantly record a high AAGR of around 7.59% for electrical energy from 28078.53MDH / year in 2013 to 97468.74MDH/year in 2030 while thermal energy marks a low decrease of about -0.3% ranging from 10836,13 MDH/year in 2013 to 10368,75MDH/year in 2030, which is presented below in figure 8. For this reasons some recommendations have been dictated for the agro-food sector and are presented below:

- Installation of economic lamps
- Optimization of compressed air
- Installation of thermal solar panels for process water heating
- Installation of energy management system
- Installation of photovoltaic panels
- Improvement of natural lighting
- Optimization of refrigeration units

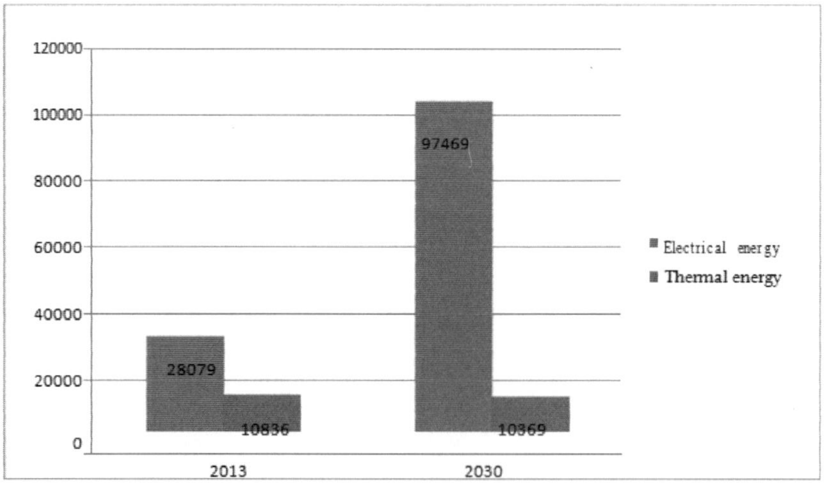

Figure 8. Moroccan energy bill in MDH of 2013 as well as for the 2030

Acknowledgments
This work was done in the laboratory" Applied Thermodynamics and Solid Combustibles (ATSC)", Mohammedia School of Engineers (EMI), Rabat- Morocco, under the direction of professor Doctor Abdellatif TOUZANI .

References

[1] Moroccan Industry Observatory, 2013

[2] D. A. Lahbabi et P. D. A. Touzani, «État des lieux : les énergies renouvelables et l'efficacité énergétique dans les secteurs de l'agriculture et de l'agro-alimentaire au Maroc : applications technologiques, conditions cadres et opportunités d'emploi- page : 120,» Deutsche Gesellschaft für Internationale Zusammenarbeit (GIZ) GmbH, 2016

[3] Moroccan Direction of the Observation of the Cooperation and Communication (DOCC), 2013

[4] D. A. Lahbabi et P. D. A. Touzani, «État des lieux : les énergies renouvelables et l'efficacité énergétique dans les secteurs de l'agriculture et de l'agro-alimentaire au Maroc : applications technologiques, conditions cadres et opportunités d'emploi- page : 146,» Deutsche Gesellschaft für Internationale Zusammenarbeit (GIZ) GmbH, 2016.

[5] "Office National Interprofessionnel des Céreales et des Légumineuses." [Online]. Available: http://www.onicl.org.ma/Principale.aspx?Module=3. [Accessed: 15-Apr-2016].

[6] "Les chiffres clés de la filière céréales et légumineuses - Fellah Trade." [Online]. Available: http://www.fellah-trade.com/fr/info-filiere/chiffres-cles/cereales-et-legumineuses. [Accessed: 14-Apr-2016].

[7] "Filière Oléicole | Département de l'agriculture - Ministère de l'agriculture et de la pêche maritime." [Online]. Available: http://www.agriculture.gov.ma/pages/acces-fillieres/filiere-oleicole. [Accessed: 11-Apr-2016].

[8] "Chiffres clés de l'aquaculture - Info filière - Fellah Trade." [Online]. Available: http://www.fellah-trade.com/fr/info-filiere/chiffres-cles/peche. [Accessed: 15-Apr-2016].

[9] Reference Document on Best Available Techniques in the Food, Drink and Milk industries, 2006

IOP Conf. Series: Materials Science and Engineering **212** (2017) 012024 doi:10.1088/1757-899X/212/1/012024

Internal Logistics System Selection with Total Cost of Ownership Analysis

Inês Araújo[1], Carina Pimentel[2], Radu Godina[3], João C. O. Matias[2,3]

[1]Department of Economics Management, Industrial Engineering and Tourism, Universidade de Aveiro, Campus Universitário de Santiago, 3810-193, Aveiro (Portugal).

[2]GOVCOPP& Department of Economics Management, Industrial Engineering and Tourism, Universidade de Aveiro, Campus Universitário de Santiago, 3810-193, Aveiro (Portugal).

[3]C-MAST, Department of Electromechanical Engineering, University of Beira Interior. Edifício I das Engenharias. Calçada Fonte do Lameiro, 1. P-6200-001 Covilhã (Portugal).

ines.araujo@ua.pt; carina.pimentel@ua.pt;rd@ubi.pt ; jmatias@ua.pt

Abstract. In this paper a methodology was followed in order to support the decision-making of one industrial unit regarding its internal logistics system. The addressed factory was facing issues with their internal logistics approach.Some alternatives were pointed out and a proper total cost of ownership (TCO) analysis was developed. This analysis was taken in order to demonstrate the more cost-effective solution for the internal logistics system. This tool is more and more valued by the companies, due to their willing to reduce the costs that are associated with the way of doing business. Despite the proposal of the best choice for the internal logistics system of the enterprise, this study also intends to present some conclusions about the match between the nature of the industrial unit and the logistics systems that best fit the requirements of those.

1. Introduction

Nowadays companies are facing several demanding challenges, mostly due to the globalization and the instability lived in the marketplace. In order to maintain a strong position in the market, enterprises are compelled to develop goods with higher quality at the best possible price for the consumer [1]. Thus, all enterprises share the concern of being effective in the way they do business. Therefore, cost associated with the business is a well-knowntopicto managers[2].

It has beenwell established that the enterprises can get to know which activities add value and witch do not, by developing a proper TCO analysis. By doing so, enterprises can find the processes that can be improved and ultimatelyimprove the manner the factory does business[3]. Although the TCO approach can be applied in all stages of the business, it is more desirable, to be used in the early stage of the business or project. By doing so, it is likely that the TCO will be minimized [4]. Also, it is easier at this stage to propose alternatives, if the project design exceeds the budget. In [5] is argued that the TCO concept assumes a huge role in high-tech decision-making process. The same authors also defend that the TCO is a tool that improves the quality of decisions and communication.

IOP Conf. Series: Materials Science and Engineering **212** (2017) 012024 doi:10.1088/1757-899X/212/1/012024

This paper presents an investigation carried out in an industrial company that was facing a few challenges on its internal logistics system. The company was considering two alternative solutions to its current material handling system so as to improve its internal material flows. Such solutions were considered in order to suit the company needs and were evaluated through a TCO analysis.The main objective of this study is to reach an alternative system that suits the needs of this type of industrial company, without compromising the target budget to spend in the internal logistics process with the needed resources (capital and human).

For the accomplishment of this study, a deep knowledge about the company and of its internal logistics process was needed. Thus, the researcher spent a considerable period of time in the facilities of the company. This period of time was fundamental to the identification of the constraints and specifications of the current internal logistics system.Also, it was necessary to create some connections with the company, in order to get access to data for the TCO analysis. During the study, it was imperative to know which solutions were available in the marketplace. Thus, the researcher conducted some meetings with suppliers of conveyor systems and milk run systems, in order to understand and gain the required knowledge about the alternative systems.

The paper is organized as follows. The logistics systems are described in Section 2. In section 3 the enterprise and the core business are briefly described. In Section 4 the applied methodology is discussed. Finally, conclusions are summarized in Section 5.

2. The logistics systems

The current material handling system is characterized by the traditional method. The materials are conducted from the company warehouses or from the workstations to the place where they are required, typically a workstation in the production area or a warehouse near the production space, by a stacker or a forklift. Using this method the unit load for the supply of goods (raw materials, work in progress or end items) is the pallet, resulting in several major drawbacks for the materials flows function: (1) the supplied quantities are much higher than the ones needed in the production area in a given instant time, resulting in an excess of work in process since the materials in excess remain near the workstation after the production finishing, until their fully consumption in future productions; (2) there is a lack of efficiency of the process and at the end the productivity is compromised, since the production workers are also involved in the movement of the materials, leaving for some periods of time their workstation; and (3) the principle of work design is compromised since the safety conditions offered to the workers are not, by far, the best ones due to the lack of space to perform their tasks using free body movements (the space near the workers is in general occupied by the goods supplied or to be collected after the production process).

The two alternative solutions under study are: a conveyor system and a milk run system. These solutions are more sophisticated than the materials handling method used currently by the factory unit, representing a different logistics flow philosophy. Therefore, the purpose of this study was to design the two alternative systems and to make a comparison between each other and with the current materials handling system. The comparison was performed through a TCO analysis.

2.1. The conveyors system

This solution is characterized by a combination of a belt conveyor and an overhead conveyor. On one hand, the belt conveyor is the most requested element of the family of conveyors. This occurs probably by its more attractive cost of acquisition and also by its large spectrum of applicability[6]. This equipment is ideal for transporting material in long distances and it is designed for continuous flow of material[7]. On the other hand, there is the overhead conveyor. Several different solutions are available of this type of equipment. Everysolution serves the same purpose that is transporting material overhead across the facilities, creating available area on the plant layout[6].

Since the objective of this equipment is to create the conditions for the integral dedication of the production operator to the production activities, the choice for the free conveyor is not wise. This free overhead conveyor requires the products to be moved by the action of the operator by pushing the product or in some cases by the action of gravity. Likewise, this equipment is more common in plants that have a low volume of production or have a high level of activities performed manually[6]. The solution gave by power and free is very sophisticated and requires a higher acquisition cost that is not justified for this application. Thus, the solution thought for the overhead conveyor is the powered one. This type of solution is suited to production systems with a medium to high volume of products. This equipment will also be transporting the material in a continuous movement[8] which is very fitting.

2.2. Milk Run

The philosophy that supports the principle of the milk run is a novelty that the companies are applying in order to improve their productive systems, by reducing the work in process (WIP) at the lowest level[9]. The milk run is supported by the Just-In-Time (JIT) philosophy, since the material is only transferred to the workstation when it is needed [10].This equipment is the result of the tow in combination with the carriages. Since the milk run is manually operated, there is a requirement for the allocation of a human resource to drive the equipment[11].For the accomplishment of the principle of work of this equipment, there are some key issues that must be addressed, like the design of proper routes and time periods for each route[12].Another topic is the renewal of the kind of containers that are applied for the transportation of material. For the implementation of the milk run, the containers must be smaller and the quantity of product in movement on those must also be standardized[13].

3. The enterprise and the core business

The enterprise where the study was taken on is a manufacturing company belonging to the automotive industry. The enterprise is in an initial position on the supply chain where it is inserted. As so, the products that the company offers will be transformed in the next partners of the supply chain.

Thisfactory has two major clients. On one hand there is the first tier client system representing 84% of the clients. On the other hand, there are the Original Equipment Manufacturer (OEM) clients which represent 10% of the clients. The rest (6%) are considered as other clients.

The core business of the company in study is stated as manufacturing metallic components for the automotive industry. As so, the raw materials used in the facilities are giving by sheet, tube or wire of metal. Since the company is divided in several production processes (stamping, welding, wire and tube forming), it can offer a large number of different products. The total number of different final products producedby the industrial unit is around five hundred. The plant layout is shown in Figure1. In yellow is highlighted the robot welding sector, in blue the elements warehouse, and in green the final product warehouse - areas that are elements under analysis in the internal logistics system.

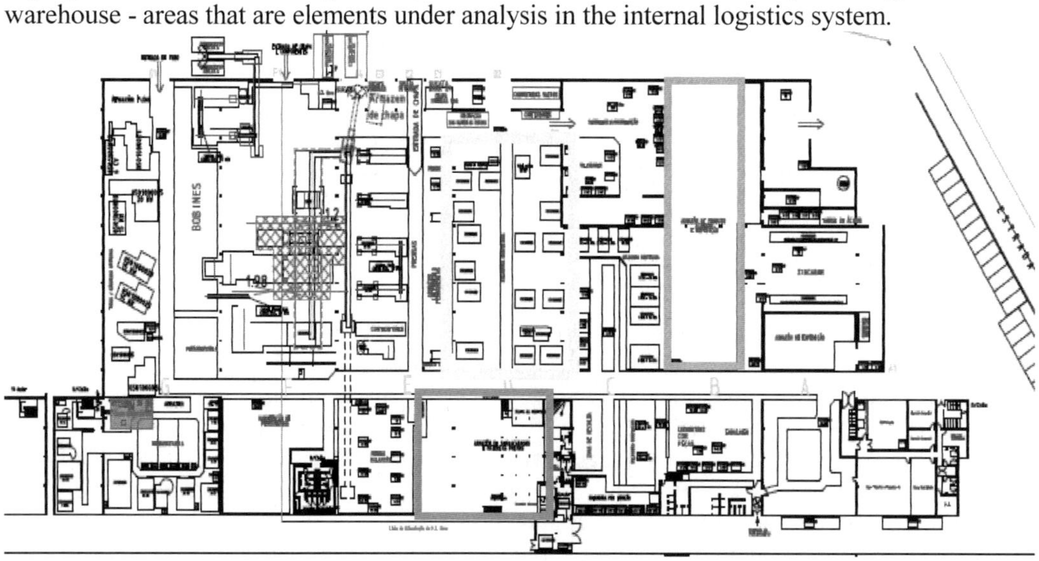

Figure 1. Layout of the factory and the highlighted areas analyzed in this study.

4. The applied methodology

The TCO analysis was developed in order to provide the more cost-efficient solution to the company. This analysis was made for a four years' period, in order to compare all the three systems (current, conveyors and milk run) in the same base. Also, it was considered that the traditional method was not implemented. As so it was necessary to purchase all the items in the three scenarios, as if it was the opening of the business.The Cost Breakdown Structure (CBS) for minor purchase was the approach followed, where the total cost is giving by acquisition, operation and maintenance costs [14]. Therefore, in each of the threecategories different costs were taken in account.

The acquisition costs were given by potential suppliers of the systems. So those values are the most accurate ones in a long-term period. The operation costs were estimated based on cost estimating relationships. Therefore, the cost associated with the non-value activities that the production operators take when they perform this logistic activity were estimated based on the man hour cost. The fuel was also estimated based on the total consumption of the forklift over a year. Specifically, it was estimated based on the distance that the forklift will do in a year, only for the pick-up of the final products of the family of products.Regarding the costs associated with maintenance, such cost was also found by cost estimating relationship. The cost of outsourcing was considered based on historical data, as well as the maintenance taken on the facilities. So the operation and maintenance cost could be less accurate in a long-term period.

It is also important to clarify that costs related to the stock were not taken into account. This assumption was made by two main reasons. In one hand, the information about the cost impact of the material stock of this family of productswas not available. In this case the consideration of cost could be completely inaccurate. On the other hand, the family of products is highly requested by the clients of the company. So the volume of production is high, and the stock of final products only occur during the week, since every Friday the product is always delivered. At the end of the week (Friday afternoon) the stock of final products is only made of the production of Friday. Nevertheless, the stock of the parts to be weld does not assume the same behavior as verified with the final products. Such parts are produced in the other plants of the company with a huge volume of production. So, these pieces parts are stored in the parts warehouse, where they are waiting to be consumed. The stock was very difficult to account, since it isn't only the plant of the warehouse that is occupied that should be addressed. There is also the cost associated with setup in the plants that produce those parts, namely the stamping sector. This is the only cost that was not taken into account. Also, as already stated in this analysis, it occurredafter the CBS for a minor purchase. On a more rigorous approach, it will be necessary to have moredetailed costs. The result of the TCO analysis is presented in the Figure 2.

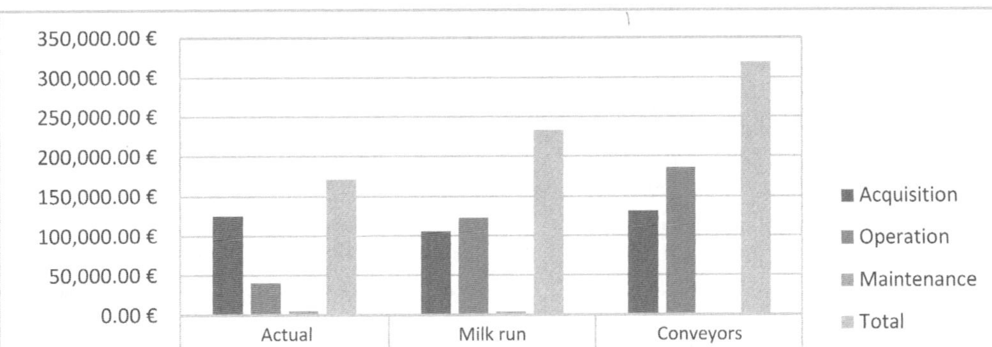

	Actual	Milk run	Conveyors
Acquisition	125,283.00 €	105,685.96 €	131,248.19 €
Operation	40,596.00 €	123,129.60 €	185,856.00 €
Maintenance	5,447.60 €	4,116.00 €	1,320.00 €
Total	171,293.60 €	232,931.56 €	318,424.19 €

Figure 2. Result of TCOanalysis.

5. Conclusion

Remarkably, the conveyors system demonstrated to be the most expensive solution in all the categories, besides the maintenance. The cost of maintenance was estimated based on the number of hours that the enterprise actually spentin maintenance of each lubrication tunnel. So this value can be underestimated, since the breakdowns are not considered. Despite the lesser cost being associated with the maintenance activities the total cost of this solution is remarkably higher than the other ones. And because of that, it won't be presented to the factory as the best solution.

The milk run has shown two categories (acquisition and maintenance) that are less expensive than the traditional method. Yet, the total cost is higher than the current method. Here is a good example that taking only the acquisition cost can lead to hasty decisions. In this context, the limitation of the milk run is in the operation cost, becausethe logistics operator is more dedicated to the activity of supply and pick-up of workstations (40 minutes each hour) than in the current system. In the current system the warehouse staff goes only once a day to the workstations, in order to supply them with the parts for the welding process.With such results, the best choice is to maintain the current system, although the factory unit must be aware that also other criteria exist besides the cost.

Acknowledgments

The current study was funded in part by *Fundação para a Ciência e Tecnologia*(FCT), under project UID/EMS/00151/2013 C-MAST, with reference POCI-01-0145-FEDER-007718.

References

[1] R. Godina, J. C. Matias, and S. G. Azevedo, "Quality Improvement With Statistical Process Control in the Automotive Industry," *Int. J. Ind. Eng. Manag.*, vol. 7, no. 1, pp. 1–8, 2016.
[2] F. A. Cummins, "Chapter 1 - The Agile Enterprise," in *Building the Agile Enterprise (Second Edition)*, Boston: Morgan Kaufmann, 2017, pp. 1–34.
[3] L. M. Ellram, "Total cost of ownership: an analysis approach for purchasing," *Int. J. Phys. Distrib. Logist. Manag.*, vol. 25, no. 8, pp. 4–23, Oct. 1995.
[4] F. Caniato, S. Ronchi, D. Luzzini, and O. Brivio, "Total cost of ownership along the supply chain: a model applied to the tinting industry," *Prod. Plan. Control*, vol. 26, no. 6, pp. 427–437, Apr. 2015.
[5] J. Heilala, K. Helin, and J. Montonen, "Total cost of ownership analysis for modular final assembly systems," *Int. J. Prod. Res.*, vol. 44, no. 18–19, pp. 3967–3988, Sep. 2006.
[6] P. M. McGuire, *Conveyors: Application, Selection, and Integration*. CRC Press, 2009.
[7] J. Hiltermann, G. Lodewijks, D. L. Schott, J. C. Rijsenbrij, J. A. J. M. Dekkers, and Y. Pang, "A Methodology to Predict Power Savings of Troughed Belt Conveyors by Speed Control," *Part. Sci. Technol.*, vol. 29, no. 1, pp. 14–27, Jan. 2011.
[8] D. He, Y. Pang, and G. Lodewijks, "Green operations of belt conveyors by means of speed control," *Appl. Energy*, vol. 188, pp. 330–341, Feb. 2017.
[9] T. Nemoto and W. Rothengatter, "Efficient Green Logistics in Urban Areas: Milk Run Logistics in the Automotive Industry," in *Sustainable Transport for Chinese Cities*, vol. 3, Emerald Group Publishing Limited, 2012, pp. 319–337.
[10] D. Gyulai, A. Pfeiffer, T. Sobottka, and J. Váncza, "Milkrun Vehicle Routing Approach for Shop-floor Logistics," *Procedia CIRP*, vol. 7, pp. 127–132, Jan. 2013.
[11] E. Klenk, S. Galka, and W. A. Günthner, "Operating Strategies for In-Plant Milk-Run Systems," *IFAC-Pap.*, vol. 48, no. 3, pp. 1882–1887, Jan. 2015.
[12] H. S. Kilic, M. B. Durmusoglu, and M. Baskak, "Classification and modeling for in-plant milk-run distribution systems," *Int. J. Adv. Manuf. Technol.*, vol. 62, no. 9–12, pp. 1135–1146, Oct. 2012.
[13] P. Olesen, D. Powell, H.-H. Hvolby, and K. Fraser, "Using lean principles to drive operational improvements in intermodal container facilities: A conceptual framework," *J. Facil. Manag.*, vol. 13, no. 3, pp. 266–281, Jun. 2015.
[14] V. Ebrahimipour, B. Maleki Shoja, and S. Li, "Supplier selection considering product structure and product life cycle cost," *Int. J. Qual. Reliab. Manag.*, vol. 33, no. 5, pp. 654–675, Apr. 2016.

IOP Conf. Series: Materials Science and Engineering **212** (2017) 012025 doi:10.1088/1757-899X/212/1/012025

Unified Information Access in Product Creation with an Integrated Control Desk

Kevin Wrasse[1], Holger Diener[2], Haygazun Hayka[1], Rainer Stark[1,3]

[1] Fraunhofer Institute for Production Systems and Design Technology (IPK), Berlin, Germany

[2] Fraunhofer Institute for Computer Graphics Research IGD, Rostock, Germany

[3]Technical University Berlin, Germany

kevin.robert.wrasse@ipk.fraunhofer.de

Abstract. Customersdemand for individualized products leads to a large variety of different products in small series and single-unit production. A high flexibility pressure in product creation is one result of this trend. In order to counteract the pressure, the information steadily increasing by Industry4.0 must be made available at the workplace. Additionally, a better exchange of information between product development, production planning and production is necessary. The improvement of individual systems, like CAD, PDM, ERP and MES, can only achieve this to a limited extent. Since they mostly use systems from different manufacturers, the necessary deeperintegration ofinformation is only feasible for SMEsto a limited extend. The presented control desk helps to ensure a more flexible product creation as well as information exchange. It captures information from different IT systems in the production process and presents them integrated, task-oriented and oriented to the user's mental model, e.g. information of the production combined with the 3D model of product parts, or information about product development on the 3D model of the production. The solution is a digital 3D model of the manufacturing environment, which is enriched by billboards for a quick information overview and web service windows to access detailed MES and PDM information.By this, the level of abstraction can be reduced and reacts to changed requirements in the short term, making informed decisions. The interaction with the control stands utilizes the touch skills of mobile and fixed systems such as smartphones, tablets and multitouch tables.

1. Introduction

The evolution in manufacturing under the term "Industry 4.0" aims at further automation by so-called cyber-physical systems in order to make production processes more efficient. This leads to an increasinginvestment of companies in specialized IT systems for digital product development and production. Nevertheless, there are always areas of product creation in which human beings play an important role. Complex heterogeneous IT systems can lead to a high cognitive load and ultimately to errors. An integrating control desk for customized information supply was developed to decrease this problem.Previous work for information supply was the project T3PADby the Technical UniversityBraunschweigin which a multi-table table for factory planning was developed. This provided only a 2D plan view of the factory as well as various libraries with virtual objects. Again, no further integration of information from other areas was planned.[1] The SCA Multitouch Prototype from ABB and the Interactive Institute Umeå provide up-to-date production figures based on a multitouch 3D model.

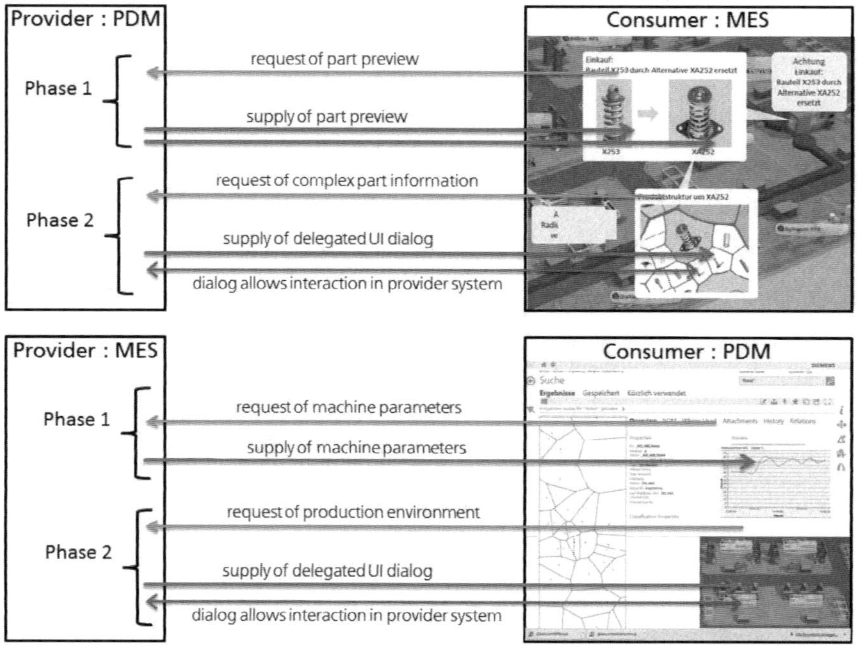

Figure 1. Information flow in control desk demonstrator

However, integrating access to other IT systems is also missing here. The system is only a prototype and is not available as a product.[2] The company iconics distributes Genesis64 for the visualization of company data in 3D models. The software combines operational data of the company but does not support product planning and does not allow any exchange of information.[3]Each of the described systems are only takinglimited aspects of product development into account. The need for a cross-sectoral exchange of information is not covered.In product creation, the processes of product development and production have been largely separated from one another. This is also reflected in the software tools for both processes. An exception here are the CAD-CAM process chains, in which the results of the product development are transmitted directly to the production. However, in most cases only unidirectional information flow is possible.[4]This is countered by the ever-growing demand for information, as well as the demand for increasingly individualized products. This is accelerated both by the technological development of sensor technology and data processing in production, as well as the increasing use of cyber-physical systems. Despite these large quantities of information and information requirements, there is a lack of mutual availability of information between the processes of product development and production [5]. For example, the production planner does not receive any feedback on how and whether his developed models have been implemented. Producing companies need fast and easy access to all relevant product creation information. The new work area "control desks for the product creation" addresses this demand. The information from the different systems is integrated into a control desk application, which allows a number of applications (as individual workplace and for cooperative work) through expansion modules.

2. Development of use-cases

Four applications were developed and further developed during the course of the project.

2.1. Information of production for product development and production planning

The application case shows how information from production can flow into product development.During the development of a new product, feedback from the production of the previous model, as well as user feedback are incorporated and have influence on the design of the product. Employees from product development and production planning compare the new product structure with the old one and then analyze the influence of the new structure on the production. In doing so, they access information from previous production processes.Important in this application is the possibility to generate the flow of information from production back into product development and production planning. This is shown in the lower diagram of figure 1. An integrated visualization between product development and production can provide the necessary attention for problems.

2.2. Information of product development for manufacturing

If parts of the product have to be replaced during production or if the production process changed, the production personnel needs product information on the new components. Instead of obtaining the entire product structure as a table, the essential information is directly displayed in a 3D model of the production plant. The assignment of parts to manufacturing machines or to the production section can be recognized directly.If necessary, the worker is provided with the entire product structure in order to obtain information on the context of the product part. This is shown in the upper diagram of figure 1. A few interactions can also be used to retrieve parts lists or work instructions. In order to facilitate on-site access in the production plant, the operation is designed for mobile devices or touch interaction in order not to force the employee to go to a normal PC workstation.

2.3. Planning and virtual inspection

Simulation data from the product development can also show on the model of the manufacturing plant how the processes will look after the change. All the digital planning documents of the production plant and of the product to be produced can be used as required. One example of this is an ergonomic study at individual workstations, which are considered to be product-oriented, as they are determined by the sequence of the assembly and the placement of the components.This is done by relating information from real workstations to the simulation data of the virtual workstation.Information about the real counterpart of the manufacturing plant is displayed for the operator. The 3D model of the production plant also allows a virtual inspection of the new production.

2.4. Monitoring and control of a production plant

Within a production plant, many different IT systems are used. The ERP system distributes orders to machine groups. The MES distributes these orders to individual machines. In the event of an error, the employee needs access to all systems in order to be able to react quickly and in a manner appropriate to their needs: How important is the order (customer information, time bar from ERP)? What machines are affected and which machines still have capacity (MES)? Unfortunately, the systems often have different user interfaces and different access requirements. For this reason, the control desk bundles all information on production. For a quick understanding of the displayed data, the visualization is carried out on a realistic 3D model of the production line. The production managers always receive an up-to-date overview of all key figures and operating data. Problems and faults can be detected and eliminated earlier. In order to analyze the causes of problems, the data from the previous production processes can also be called up for comparison purposes.

3. Concept

The aim was to develop a concept for the consideration of the individual mental models in the application cases and to find a general procedure for new applications. On the other hand, the concept of a common technological basis for an integrated ELP demonstrator was developed.

3.1. Mental models

In psychology, mental models are individual explanatory approaches on the mode of action and the relationships of objects of reality. They are created bit by bit through experience and cognitive processing[6].Mental models always simplify reality. Good mental models obtain the structure of the real scene (homomorphism) and contain all relevant aspects. However, some studies show that visual and even spatial models are common.Visualizations are representations of real data and thus models that the user has to interpret. The interpretation should be easy for the users and they should have little room for misinterpretations. Therefore, the visualization model should be as similar as possible to the individual mental model of the users[7]. Discussions with production managers confirmed that manyworkers were imagining the production site as a space.

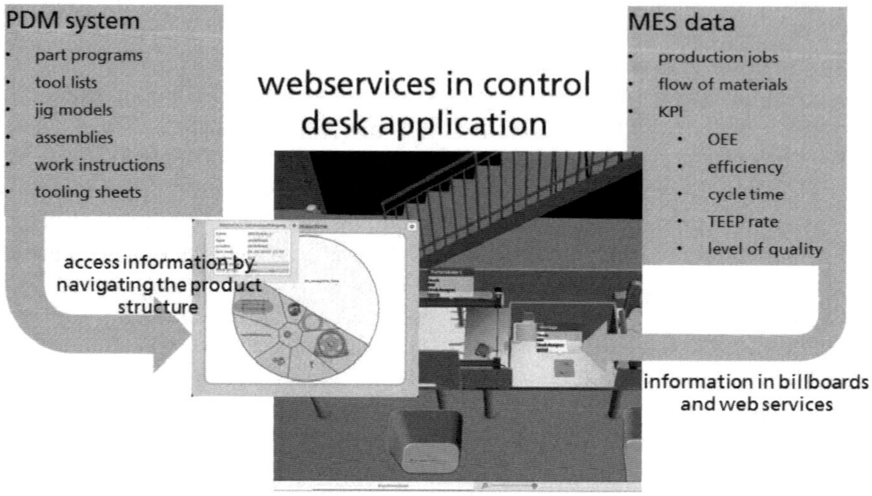

Figure 2. Information integration in control desk

For this reason, the 3D visualization of the manufacturing facility was chosen as closely as possible to the reality and the visual represen-tation of the mental model in the imagination of the workers. Similar observations have also been made in product development. Here the 3D representation of the product and its components supports the understanding of the data better than abstract tree representations. The use of 3D models and the compilation in a real-world representation of the product or production hall is a successful approach for the use of mental models. However, there is no standard of a mental model for the implementation of control desks. In the development of the control desk, the expectation of the users must be taken into account. In addition, the mental models are also to take up the usual computer work processes of the users. Users work with a set of IT tools that they are accustomed to using. If an IT tool is to integrate user groups from different areas, it is necessary for each user to recognize his mental model in the tool. This allows the user to use the information quickly and effectively. Figure 2 shows the information that is aggregated in the two integrated systems. Information from the PDM system with production planning data and live data from the production process have to be shown in a single screen.

3.2. Preparatory work

In the preliminary work very different technologies (web-based application with Java and JavaScript and desktop application with Unity3D and C #) were used. At the start of the project it was, however, foreseeable that Unity3D will be able to create web-enabled applications in the future. For this reason, various possibilities of web-based integration were investigated.Both modules offer functionalities for a particular area of product creation. Plant@Hand3D offers the possibility to model the graphic representation and user interaction in a Unity3D development environment.This allows a factory model with short information and detail windows to be added. For the integration in Plant@Hand3D an extension for the display of web pages was developed and the modules for the MES and the PLM functionalities were implemented as websitesThe second basic component - the PDM module - previously existed as a standalone Java application developed for multitouchtables[8]. For the integration, the application had to be web-enabled and almost completely rewritten in JavaScript.The access to the Siemens Teamcenter web services and to PLMXML files was taken over. The MES

IOP Conf. Series: Materials Science and Engineering **212** (2017) 012025 doi:10.1088/1757-899X/212/1/012025

functionalities were implemented in PHP and JavaScript. The allocation of the MES and PDM modules is controlled by an XML file, in which their web addresses are stored.

4. Integrated solution

After the application cases were defined, the technical concepts as well as the handling of the individual mental models were established, the refinement of the existing technological basis took place during the implementation phase.As a result, the modules were successfully integrated into one another in order to meet the requirements of the application.

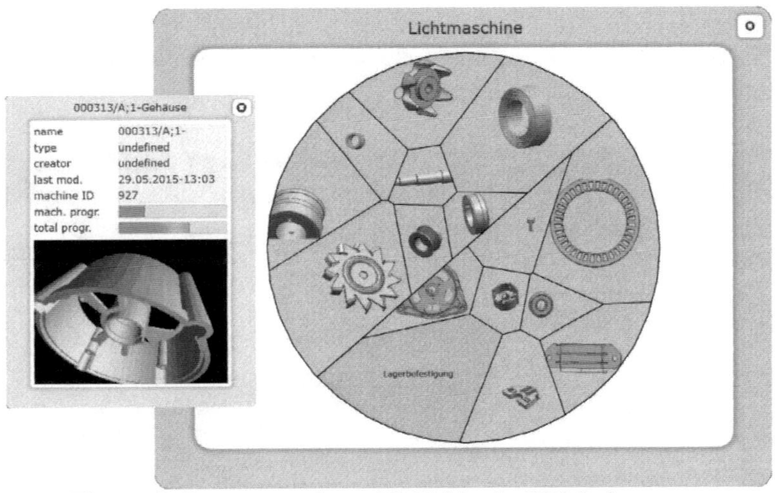

Figure 3. Representation of BOM in the PDM viewer

4.1. Components for control desks in production monitoring

All necessary elements for interaction with the control center (settings, zoom and navigation, selection of the data views and other context parameters) are displayed. The basic system is multitouch-capable and supports various end devices (PC, mobile phone, tablets) and platforms (web browser, Windows, Android, iOs). It thus achieves a very high range for various possible applications including mobile scenarios.A workflow has been created for the import of 3D models, which also allows highly detailed models to be simplified and used in the control room. An editor has been developed to reduce the development time and to simplify the customization process so that employees can also make changes without programming knowledge. The editor allows the parameterization of the individual machines to be identified in the various IT systems of the company. In this way, they can be connected to data from these systems.[9]Most data views are represented by short information and detail windows. For this purpose, some standard components were created for the visualizations. These can be used whenever both the data sources as well as the requirements of the customer permit it. As standard components, two types of line diagrams as well as a timeline and a tor- ding diagram were created to display KPI (Key Performance Indicators), machine data and status history. In addition, data views can also be directly affected by the 3D model (position, color, etc.) or as separate 3D objects in the production model. In order to visualize the flow of an order, additional 3D objects (arrows) are displayed between the machines to display the transaction data. The movement of these arrows can be written either as a planning task in the underlying planning systems or as a feedback for disturbances in the process.

4.2. Integration of the component PDM-Viewer

The PDM viewer is a single-page application for displaying product structures (see figure 3). First, it was developed as a standalone site with network access to a server of the Teamcenter PDM / PLM tool. From this server the application got the data of the product structure as well as of the individual assemblies and components using Web services and SOAP. With its integration within the framework of the MEF-ELP project, the application is no longer stand-alone, but is integrated into the Unity3D-based factory-control and monitoring program Plant@Hand3D via a web viewport. In the developed demonstrator, this variant retrieves its data both via Web Services / SOAP and a PLMXML file derived from the Teamcenter PLM system. This ensures the portability of the displayed information. This internally establishes a connection to the Siemens Teamcenter in order to be able to supply data and all structural elements, regardless of whether they are available on the network or locally. In addition, the connector controls the visualizations of the various product structures. The product structure is displayed using a Voronoi diagram. In this diagram, a cell represents a component or assembly, with the cell of an assembly being able to "go in" to open the underlying plane again in a subordinate Voronoi diagram.

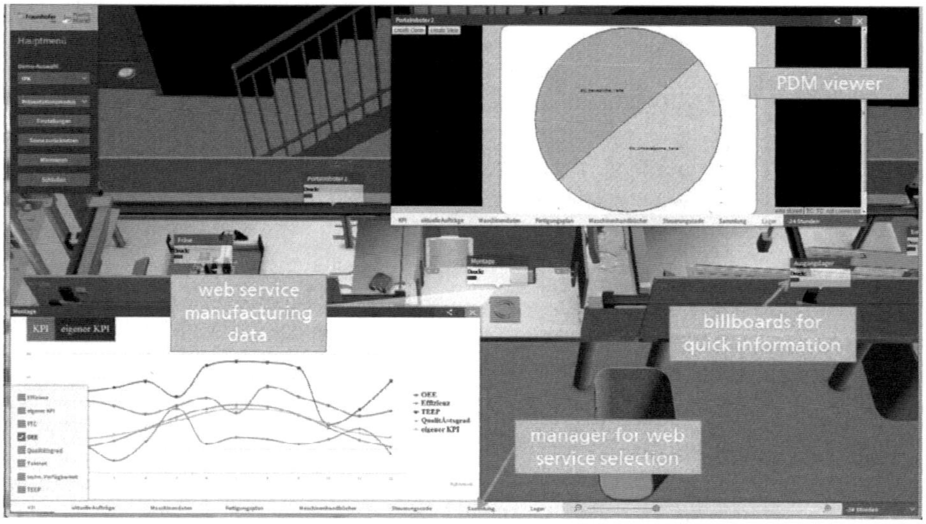

Figure 4. Integrated control desk demonstrator

Figure 4 shows the embedding of the module in the application cases. At first, a circular representation of the product structure appears in the Voronoidiagram[10]. The layers of the product structure are represented by the resolution levels in the tiles. As the top level, the entire system fills the circle completely.In the next detail stage the circle is subdivided into the subsystems. The figure shows an alternator, which is divided into the two subsystems rotor and stator. The next detail shows the components that make up the subsystems. Navigating to the respective detail levels is done by means of a pop-up menu, which, if pressed for a longer time, opens onto the Voronoi diagram.Another button opens an information dialog for the component. The existing metadata from the PDM system is displayed first.These are, for example, the item ID, the creator, and the last change. The current location of the component as well as the progress of the production can be displayed to a specific component of an order. The information is supplemented by a 3D view of the component, which can be rotated and enlarged with touch gestures. The lines of the information dialog are modularly defined in the program code and can be arbitrary extended or changed. In this way, this tool is easily adaptable to new applications. The new implementation has made it possible to easily integrate the PDM viewer as a component in the control desk, thus providing information from product planning.

4.3. Implementation of the component »MES-Webservices«

The machine data is processed and displayed in a highchart diagram in a javascript file. In addition, the navigation menu can be used to access a dashboard view, which is linked to another javascript file (see figure 4). The dashboard view provides an overview of machine parameters and compares them so that a quick overview can be obtained. In the current implementation, the data are obtained from a MONGO database, which is well suited for processing large amounts of data. The data can be simply integrated into the database from XML or JSON files and displayed in the dashboard.

4.4. Integration of different IT systems

The solution should make it possible to link the various data together, but not to duplicate data, but only to retrieve data from the data-carrying systems. For this purpose, a software architecture (see Fig. 5) has been developed which can collect and display data from different systems via different interfaces. All data can be displayed via different data views (managers). There are generally two stages. In the first stage, information is displayed as short information directly on the machine. If further details are required, a detail window can be opened from each brief information as a second stage. Data sources are generally accessed via REST or SOAP Web services. Direct access to a database is also possible. The frequency of the data queries can be scaled as required by the data record. The data context (time, type of data, etc.) can be set, and workflows in the IT systems can also be triggered (fault messages, rescheduling, etc.).Some companies use fully web-based IT systems that run the client application in the browser. For these cases a webrendering component was developed so that the contents of the web

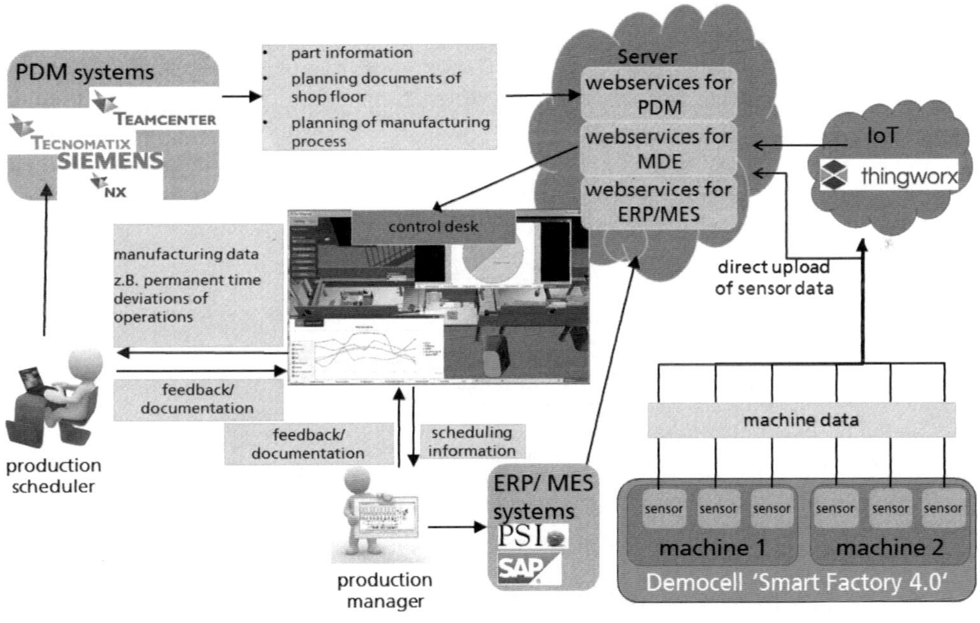

Figure 5. Architecture of IT system integration.

application can be executed directly in detail windows of the control desk. The integration of individual web pages or entire web applications is also completely possible without programming. A configuration file, which defines the URLs and additional parameters for defining the context, allows integration in a few minutes.

5. Evaluation

Within the scope of the project, the technical implementation of the individual components was dividedbetween the partners. The integration of the components via a web rendering component or the direct integration in the browser broughtsuitable results. In the course of the conversions, integration tests were carried out repeatedly with parts of the components and the respective state of the control room applications, so that a technical evaluation of the components was always present.As an expert evaluation, components and integrated demonstrators were repeatedly discussed with various industrial partners within the project. The most striking was the scenario of product development and production monitoring. At the moment, the companies still felt a small need for an increased interlocking of PDM and MES information, as the data situation in the companies is not yet sufficient. In order to present convincing examples for the visualization of information from the production, the control station was integrated and tested at the Fraunhofer IPK on the SmartFactory 4.0 demonstrator.The possibilities of the integrated control panel were presented to users in the PTK conference in 2017. In the demonstrator, the use-cases from chapter 2 are implemented to point out advantages for manufacturing. The cooperation with industrial partners in this field is important, since the interlinking of product development, production planning and production will increase in importance in the following years due to the growing importance of digitalization.

6. Conclusion

With this demonstrator of an integrated control desk, it is possible to apply and test various use cases of production processes. Information from two distinct processes of product creation were integrated in one tool. The visualization concept allows an intuitive navigation and access on relevant information. It is based on different mental models and the visual integration of data from the different areas supports the cooperation between engineers of the various work areas. The application provides a uniform interface for querying information from the product development and production planning and control systems, for monitoring data, and for initiating workflows. In particular, through the increase in information and the demands on the flexibility of work processes, an integrated and mobile-accessible presentation can achieve a better information supply. The solution is aimed at small and medium-sized companies in the mechanical and plant-engineering sector as well as suppliers for automotive manufacturers and aviation. Many of these companies are in global competition and have to switch to efficient, individualized production. Control desks help these companies with their information from the various IT systems more efficiently.

7. Acknowledgement

The project “Raise efficiency by control desks in product creation“ was funded by the Fraunhofer Society in the scope of an MFE project. MFE projects are funded to elaborate solutions that support SMEs in Germany. The project was conducted in cooperation by the Fraunhofer IPK and Fraunhofer IGD-R.

References

[1] Dombrowski, U.; Hennersdorf, S. (2009): Business process models supporting participatory layout planning. In: 2009 International Conference on Intelligent Engineering Systems (INES). Barbados, S. 113–117.

[2] Zarin, Ru (2011): ProjectSCA Multitouch Prototype. RISE Interactive Umeå. Sweden. Online available https://www.tii.se/projects/LEAN#description.

[3] Agrusa, Ross (2017): GENESIS64: Next Generation in HMI/SCADA Automation Software. ICONICS, Inc. Online available http://www.iconics.com/Home/Products/HMI-SCADA/GENESIS64.aspx#.WKrl3n-i5P1.

[4] Jianxin, Xu; Jingtao, Zhou; Jianfeng, Li; Siyang, Dong (2010): 3D-Model-Based Integrated Capp System for Aircraft Structural Parts. In: George Q. Huang, K. L. Mak und Paul G. Maropoulos (Hg.): Proceedings of the 6th CIRP-Sponsored International Conference on Digital Enterprise Technology. Berlin, Heidelberg: Springer Berlin Heidelberg, S. 15–25.

[5] Xu, Fei (2009): System Integration and Application of PDM CAPP and MES Based on Multi-views Management of BOM. In: 2009 Second International Workshop on Computer Science

and Engineering. Qingdao, China, S. 200–203.

[6] Rauh, Reinhold; Knauff, Markus; Schlieder, Christoph (1998): Leitthema: Räumliche mentale Modelle/Spatial Mental Models. In: Kognitionswissenschaft 7 (1), S. 1–4. DOI: 10.1007/BF03354956.

[7] Wrasse, Kevin; Brandenburg, Elisabeth; Hayka, Haygazun; Stark, Rainer (2015): Visualisierung von Anforderungen an physische Prototypen mit Augmented Reality. In: Jürgen Gausemeier, Michael Grafe und Meyer auf der Heide, Friedhelm (Hg.): Augmented & Virtual Reality in der Produktentstehung. Paderborn: Heinz Nixdorf Institut Universität Paderborn (Verlagsschriftenreihe des Heinz Nixdorf Instituts, Band 342).

[8] Wolter, Lars; Woll, Robert: Kollaborativer Design Review am Multitouchtisch. In: Reiterer (Hg.) 2012 – Interaktiv informiert, S. 439–443.

[9] Aehnelt, M. (2016): Plant@Hand - Intelligente Assistenzsysteme für Industrie 4.0. In: wt Werkstattstechnik 4-2016. Springer VDI-Verlag, Düsseldorf, 2016

[10] Schulze, E.-E.; Wolter, L.; Hayka, H.; Röhlig, M. (2012): Intuitive Interaktion mit Strukturdaten aus einem PLM-System. In: Ralph Stelzer, Karl-Heinrich Grote, Klaus Brökel, Frank Rieg und Jörg Feldhusen (Hg.): Entwerfen, entwickeln, erleben. Methoden und Werkzeuge in der Produktenentwicklung. Dresden: TUDpress, S. 71–87.

IOP Conf. Series: Materials Science and Engineering **212** (2017) 012026 doi:10.1088/1757-899X/212/1/012026

Grey Comprehensive Evaluation of Biomass Power Generation Project Based on Group Judgement

Huicong Xia, Dongxiao Niu

Economics and Management School/ North China Electric Power University, Beijing, China

1656184583@qq.com, niudx@126.com

Abstract. The comprehensive evaluation of benefit is an important task needed to be carried out at all stages of biomass power generation projects. This paper proposed an improved grey comprehensive evaluation method based on triangle whiten function. To improve the objectivity of weight calculation result of only reference comparison judgment method, this paper introduced group judgment to the weighting process. In the process of grey comprehensive evaluation, this paper invited a number of experts to estimate the benefit level of projects, and optimized the basic estimations based on the minimum variance principle to improve the accuracy of evaluation result. Taking a biomass power generation project as an example, the grey comprehensive evaluation result showed that the benefit level of this project was good. This example demonstrates the feasibility of grey comprehensive evaluation method based on group judgment for benefit evaluation of biomass power generation project.

1. Introduction

With the continuous consumption of fossil fuels such as coal, oil and natural gas, and the increasing environmental problems caused by fossil energy over-burning, the attention on renewable and clean energy is increasing. As a kind of renewable energy, biomass energy is stable, clean and safe, and the rational and efficient use of biomass energy has become an important issue for the development of current society. Biomass power generation is a major form of biomass energy utilization, including biomass direct combustion power generation, biomass mixed combustion power generation, biomass gasification power generation, methane electricity generation and so on [1]. As of 2015, the global installed capacity of biomass power is about 100 million kilowatts, of which the United States holds 15.9 million kilowatts and Brazil holds 11 million kilowatts. Biomass cogeneration has become an important heating mode in Europe, especially in the Nordic countries. China is rich in biomass resources, the national biomass resources which can be used as energy sources is approximately equivalent to 460 million tons of standard coal per year. As of 2015, the total installed capacity of biomass power generation in China is about 10.3 million kilowatts, of which agriculture and forest biomass direct combustion power generation holds about 5.3 million kilowatts, waste incineration power generation holds about 4.7 million kilowatts, and methane electricity generation holds about 0.3 million kilowatts, and the annual generated energy is about 52 billion kWh [2]. Biomass power generation has good environmental and social impact, but the deviation of actual economic benefits from expectations of it is also a troubling question, and it causes some of the environmental problems and food security issues [3]. Therefore, it is necessary to establish a comprehensive evaluation method, which takes the investment return, environmental impact and social benefit as the basis for judging the feasibility and economy of biomass power generation projects.

The comprehensive evaluation index system of biomass power generation projects is usually established from the aspects of management level, technical effect, economic benefit, environmental impact and social benefit, while economic benefit, environmental impact and social benefit are the main analysis objects [4,5]. The commonly used comprehensive evaluation methods include PCA (principal component analysis), intelligent comprehensive evaluation method, fuzzy comprehensive evaluation method, and grey comprehensive evaluation method [6]. The main principle of PCA is dimension reduction. To simplify the comprehensive process, PCA analyzes the internal relationship between indicators based on multiple sets of data with same attributes, and transforms the complex indicators into an independent comprehensive index system which contains most of the information of the original indicators. Intelligent comprehensive evaluation method establishes a mapping from indicators to evaluation results through a lot of training and learning. The premise of using these two methods is that the number of comparable samples is large enough, but for biomass power generation projects, it is difficult to obtain sufficient comparable data, and therefore PCA and intelligent comprehensive evaluation method do not apply. Fuzzy comprehensive evaluation method and grey comprehensive evaluation method can be used for the evaluation of a single object, but the subjective degree of fuzzy comprehensive evaluation method is too high as experts directly determine the level of indicators, so this paper selects the grey comprehensive evaluation method based on triangle whiten function. The triangle whiten function describes the subjection degree of a value to a value range. Dai Ran uses the triangle whiten function to evaluate a hydraulic engineering project, and proves the accuracy and reliability of the model [7]. Qing Song, Wang Ye, and Peng Peng use the grey whitening weight function to evaluate the security risk of power system, and prove the feasibility and practicability of grey clustering method [8]. Liu Sifeng, Xie Naiming put forward the grey evaluation method based on center-point triangular whitenization weight function, and prove the validity of the method with an example [9].

The only reference comparison judgment method for weight calculation and the grey comprehensive evaluation method for evaluation are greatly influenced by subjective factors, so this paper applies group judgment to these two steps at the same time.

2. Grey comprehensive evaluation model of biomass power generation project based on group judgement

2.1. The establishment of evaluation index system

Evaluation index system is the link between evaluation method and evaluation object. Scientific and reasonable evaluation index system plays an important role in making a fair evaluation conclusion. In general, economic benefit is a core issue for investors, and an important criterion for operation level assessment. Due to the unique social publicity and environmental protection property of biomass power generation projects, the state and society attach great importance to its social and environmental benefit. In order to ensure the comprehensiveness of the evaluation process and the reliability of the evaluation result, the evaluation indicators can be divided into economic benefit, environmental benefit and social benefit. The index system is shown in Table 1.

Table 1. Comprehensive evaluation index system of biomass power generation project

Target layer	Criterion layer	Index layer	Index symbol
Comprehensive benefit evaluation of biomass power generation project	Economic benefit	Net present value	A_1
		Internal rate of return	A_2
		Payback period	A_3
		Return on total assets	A_4
	Social benefit	Improving the awareness of environmental protection	B_1

		Providing employment opportunities	B_2
		Increasing farmers' income	B_3
	Environmental benefit	SO_2, CO_2 emissions	C_1
		Dust effect	C_2
		Noise effect	C_3
		Wastewater impact	C_4

The selected economic benefit indicators reflect the investment value, investment efficiency and profitability of projects. The selection principle of social benefit indicators is to reflect the social environment contribution and social economy contribution of projects. Environmental benefit indicators mainly reflect the air pollution, dust pollution, noise pollution and wastewater discharge.

2.2. *The weight calculation based on only reference comparison judgment method*

In practical problems, it is difficult for experts to give an only and accurate relative importance estimation of two indicators, while it is relatively easy to use a certain value range to express their subjective understanding. Therefore, in this paper, experts use interval numbers to estimate their subjective understanding of relative importance between indicators.

The only reference comparison judgment method first selects the least important indicator, and then uses the relative importance of residual indicators to the least important one to calculate the weight of all indicators. Interval number $D=[d_1,d_2]$ ($1\le d_1\le d_2\le 9$) is used for the estimation work, where 1 for equally important and 9 for extremely important. The least important indicator is expressed by $D=1$. The width of interval number is

$$L(D)=d_2-d_1 \tag{1}$$

The midpoint of interval number is

$$M(D)=\frac{d_1+d_2}{2}. \tag{2}$$

The interval mapping function with experts' risk attitude is

$$\varphi_\varepsilon(D)=M(D)+\varepsilon\times L(D). \tag{3}$$

ε is the risk attitude factor: $-\frac{1}{2}\le\varepsilon\le 0$ for conservative type; $\varepsilon=0$ for neutralities; $0\le\varepsilon\le\frac{1}{2}$ for adventurers.

When there is only one expert, the weight of each indicator is:

$$w_i=\frac{\varphi_\varepsilon(D_i)}{\sum_{l=1}^{m}\varphi_\varepsilon(D_l)}. \tag{4}$$

For the situation of L ($L>1$) experts:

(1) Experts have the same qualitative judgment.

When L experts have the same judgment of the least important indicator x_{jm} of the index system $\{x_j\}$, let the interval estimation of the relative importance a_i of indicator x_{ji} and x_{jm} from expert k be $D_i^{(k)}=[d_{1i}^{(k)},d_{2i}^{(k)}],(i=1,2,\cdots m;k=1,2,\cdots L)$.

Make the following treatment:

$$D_i=\bigcap_{k=1}^{L}D_i^{(k)}=[d_{1i},d_{2i}]. \tag{5}$$

Where

$$d_{1i} = \max_k\{d_{1i}^{(k)}\}, d_{2i} = \min_k\{d_{2i}^{(k)}\}. \tag{6}$$

The average blindness of the interval number a_i of L experts can be calculated as follows:

$$R_i = \frac{1}{L}\sum_{k=1}^{L}[(d_{1i} - d_{1i}^{(k)}) + (d_{2i}^{(k)} - d_{2i})], i = 1,2,...,m. \tag{7}$$

$$0 \le R_i \le d_{1i} - d_{2i} + d_{2i}^* - d_{1i}^*.$$

Where

$$d_{1i}^* = \min_k\{d_{1i}^{(k)}\}, d_{2i}^* = \max_k\{d_{2i}^{(k)}\}. \tag{8}$$

Make the following treatment: $D_i^* = [d_{1i}^*, d_{2i}^*]$. Calculate parameters:

$$\lambda_i = \frac{1}{2L}\sum_{k=1}^{L}\frac{n(D_i^{(k)})}{n(D_i^*)}, i = 1,2,...,m. \tag{9}$$

Assuming that:

$$\varphi_{\lambda_i}(D_i) = n(D_i) + \lambda_i e(D_i). \tag{10}$$

The weight of each indicator can be calculated as follows:

$$w_i = \frac{\varphi_{\lambda_i}(D_i)}{\sum_{l=1}^{m}\varphi_{\lambda_l}(D_l)}, i = 1,2,...,m. \tag{11}$$

(2) Experts have different qualitative judgment.

When L experts have different idea of the least important indicator of the index system $\{x_j\}$, there are two cases.

For the L_0 experts who have the same qualitative judgment of the least important indicator, the weight w_i^1 can be analyzed by the above steps.

For the remaining $L - L_0$ experts with different views:

$$w_i^2 = \frac{1}{L - L_0}\sum_{*k=1}^{L-L_0} w_i^{(*k)}, i = 1,2,\cdots m. \tag{12}$$

The comprehensive weight can be calculated by the next step:

$$w_i = k_1 \times w_i^1 + k_2 \times w_i^2. \tag{13}$$

Where $k_1 = \frac{L_0}{L}, k_2 = \frac{L - L_0}{L}$.

2.3. Grey comprehensive evaluation of biomass power generation project based on triangle whiten function

Whitenization weight function describes the "preference degree " of a grey number for values within different value ranges. Continuous function with determinate starting and end point, increasing function on the left side, and decreasing function on the right side, is called the typical whitenization weight function, as shown in Figure 1.

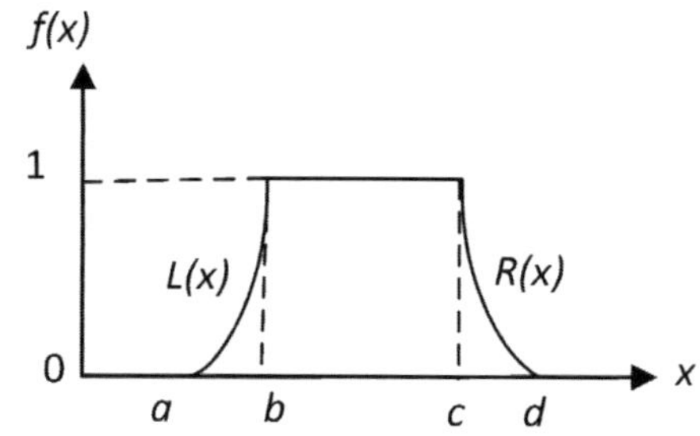

Figure 1. Typical whitenization weight function

In practical applications, in order to facilitate programming and computing, $L(x)$ and $R(x)$ are often simplified to linear function. When $L(x)$ and $R(x)$ in Figure 1 use linear function, and b , c merge into one point, whitenization weight function turns into triangle whiten function $f(x)$, as shown in Figure 2.

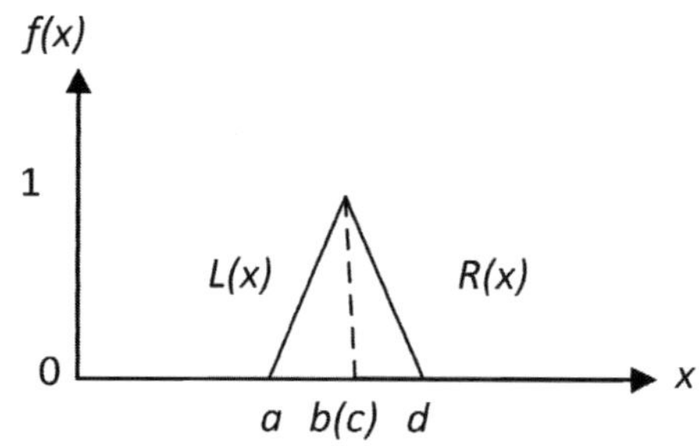

Figure 2. Triangle whiten function

Assuming that there are m indicators, s grey categories, L invited experts. The sample observation of evaluation object on the indicator j provided by expert i is $x_j^{(i)}$, $i=1,2,3,\ldots,L$, $j=1,2,\ldots,m$. The specific comprehensive evaluation steps based on $x_j^{(i)}$ are as follows:

(1) Determine the evaluation grey classifications, and select the threshold

According to the requirement of comprehensive evaluation of biomass power generation projects, determine the number of evaluation grey classifications to be s . Combine the realities of biomass power generation projects with similar experiences to determine the value range of indicators. The value range of each indicator is divided into s groups. For example, the value range $[a_{j,1},\ a_{j,s+1}]$ of indicator j is divided into $[a_{j,1},\ a_{j,2}]$, …, $[a_{j,k},\ a_{j,k+1}]$, …, $[a_{j,s},\ a_{j,s+1}]$. The threshold $a_{j,k}$ can be determined according to the actual requirement or related research result.

(2) Optimize observations

In order to improve the objectivity of evaluation results, L sets of initial estimations given by experts should be optimized according to the minimum variance principle. $x_1^{(i)},x_2^{(i)},\cdots x_m^{(i)}$ is the initial estimation from expert i . Set the optimized observations being $x_1,x_2,\cdots x_m$, the variance $\sigma^2{}_j$ can be calculated by the following equation:

$$\sigma^2{}_j=\sum_{i=1}^{L}(x_j^{(i)}-x_j)^2=\sum_{i=1}^{L}(x_j^{(i)})^2+L\times(x_j)^2-2\times x_j\times\sum_{i=1}^{L}x_j^{(i)},\ j=1,2,\cdots m. \quad (14)$$

The partial derivative of the above equation is:

$$(\sigma^2{}_j)' = 2 \times L \times x_j - 2 \times \sum_{i=1}^{L} x_j^{(i)} . \quad (15)$$

When $\sigma^2{}_j$ obtains the minimum value,

$$x_j = \frac{\sum_{i=1}^{L} x_j^{(i)}}{L} \quad (16)$$

Whereby the optimized observations $x_1, x_2, \cdots x_m$ can be calculated.

(3) Establish the triangle whiten function

Set the triangle whiten function value of $\frac{a_{j,k} + a_{j,k+1}}{2}$ to the grey classification k is 1. Connect $(\frac{a_{j,k} + a_{j,k+1}}{2}, 1)$ with the starting point $a_{j,k-1}$ of grey classification $k-1$ and the end point $a_{j,k+2}$ of grey classification $k+1$, obtain the triangle whiten function $f_j^k(x_j), j = 1,2,\ldots,m; k = 1,2,\ldots,s$ of grey classification k. The triangle whiten function describes the subjection degree of a value to a value range.

$$f_j^k(x_j) = \begin{cases} 0, x_j \notin \left[a_{j,k-1}, a_{j,k+2}\right] \\ \frac{x_j - a_{j,k-1}}{\lambda_{j,k} - a_{j,k-1}}, x_j \in \left[a_{j,k-1}, \lambda_{j,k}\right] \\ \frac{a_{j,k+2} - x_j}{a_{j,k+2} - \lambda_{j,k}}, x_j \in \left[\lambda_{j,k}, a_{j,k+2}\right] \end{cases} . \quad (17)$$

Where

$$\lambda_{j,k} = \frac{a_{j,k} + a_{j,k+1}}{2} .$$

For $f_j^1(x_j)$ and $f_j^s(x_j)$, the value range can be extended to $a_{j,0}$ and $a_{j,s+2}$ according to the specific situation, as shown in Figure 3.

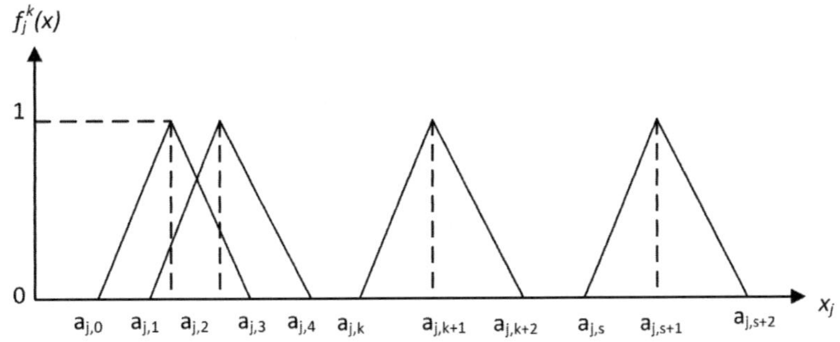

Figure 3. Triangle whiten function and its extension

(4) Calculate the comprehensive clustering coefficient

Calculate the comprehensive clustering coefficient $\sigma^k (k = 1,2,\ldots,s)$ of grey classification k :

$$\sigma^k = \sum_{j=1}^{m} f_j^k(x_j) \times w_j \,. \tag{18}$$

x_j is the optimized observation of indicator j. $f_j^k(x_j)$ is the triangle whiten function of grey classification k. w_j is the weight of indicator j.

It can be judged that the evaluation object belongs to grey classification k^* according to $\max_{1\le k\le s}\{\sigma^k\} = \sigma^{k^*}$.

3. Case study

This paper assumes that there is a garbage power generation project with two 400t / d grate incinerators and one 2×7.5MW condensing steam turbine-generator unit. Actual economic benefits are shown in Table 2.

Taking the benefit evaluation of a $1\times30MW$ biomass power generation project as an example, the project is equipped with a $130t/h$ high temperature and high-pressure biomass boiler. The remaining wheats, corn stalks and peanut shells are the main fuel, supplemented by other biomass fuels. The process water uses reclaimed water from sewage treatment plants, while domestic water, firewater and alternative water sources for production come from groundwater. Ash and slag handling system is used in this project. Actual economic benefits are shown in Table 2.

Table 2. The statistic table of actual economic benefits

Indicator (unit)	Actual value
Net present value (ten thousand yuan)	9000
Internal rate of return (%)	12.5
Payback period (year)	8.5
Return on total assets (%)	9.5

Improving the awareness of environmental protection: Biomass power generation projects can effectively solve the environmental pollution caused by traditional straw combustion, help local residents to explore clean and economic treatment of straw, and improve the environmental awareness of residents.

Providing employment opportunities: The project requires a lot of labor force during construction and operation. In addition, to a certain extent, the project can promote the development of peripheral catering trade and other sideline, and thus improve the employment situation.

Increasing farmers' income: To a certain extent, the project improves the income level of the local residents. In addition, it can help farmers to transform the biomass resources into commodity resources, form characteristic industry, promote the farmers' income and drive the local economic development.

SO_2, CO_2 emissions: The project has a spot of dust, SO_2 and NO_X emission. It uses high altitude discharge, long distance conveying, diffusion, dilution and other measures, so the impact of pollutants on the surrounding environment is small.

Dust effect: The fuel straw is piled up in a storage shed or an open-air storage yard. The top of the storage shed is closed, and the open-air storage yard is covered with canvas in the event of windy weather. Ash and slag handling system is designed to prevent dust pollution, so the dust pollution is relatively light.

Noise effect: The noise value of the project is 75 ~ 100d B (A). Boiler room, pump room and other places with large noise work independently. Induced draft fan and other machines with large rotation noise use low-noise equipment. Appropriate greening can play a positive role in dustproof, noise reduction and beautifying the environment.

Wastewater impact: The project has a sewage treatment station and a wastewater pumping station. Domestic sewage, chemical wastewater and waste water from circulation is collected by the sewage network and then discharged into the sewage treatment station. Part of the treated waste water is used for slag removal and ash removal, and the remaining part is discharged into sewage treatment plant through wastewater pumping station.

According to the performance of this project, we can evaluate its comprehensive benefits. Specific steps are as follows.

3.1. Calculate the weight of evaluation indicators

Invite 5 experts to make independent judgment on the least important indicator and give interval numbers of relative importance. The relative importance judgment is shown in Table 3.

Table 3. The result of relative importance judgment

Index symbol	$D_i^{(1)}$	$D_i^{(2)}$	$D_i^{(3)}$	$D_i^{(4)}$	$D_i^{(5)}$
A_1	[4,6]	[7,8]	[7,8]	[6,8]	[7,9]
A_2	[7,9]	[6,8]	[6,8]	[6,9]	[7,9]
A_3	[6,8]	[6,8]	[4,6]	[5,6]	[4,6]
A_4	[4,6]	[4,5]	[4,6]	[5,6]	[4,6]
B_1	[2,4]	1	1	1	1
B_2	1	[2,4]	[2,4]	[3,5]	[2,5]
B_3	[6,8]	[4,6]	[3,5]	[4,5]	[3,5]
C_1	[7,9]	[7,9]	[6,8]	[6,9]	[6,8]
C_2	[6,8]	[6,8]	[4,6]	[5,6]	[4,6]
C_3	[2,4]	[2,4]	[3,5]	[3,4]	[2,4]
C_4	[3,5]	[3,5]	[3,5]	[4,6]	[4,5]

Calculate D_i and D_i^* for the latter four experts with same qualitative judgment on the least important factor, the result is shown in table 4.

Table 4. The calculation result of D_i and D_i^*

Index symbol	D_i	D_i^*
A_1	[7,8]	[6,9]
A_2	[7,8]	[6,9]
A_3	[6,6]	[4,8]
A_4	[5,5]	[4,6]
B_1	[1,1]	[1,1]
B_2	[3,4]	[2,5]
B_3	[4,5]	[3,6]
C_1	[7,8]	[6,9]
C_2	[6,6]	[4,8]
C_3	[3,4]	[2,5]
C_4	[4,5]	[3,6]

Combine Eq. 9 with Eq.11 to calculate the weight opinion w_i^1 of the latter four experts. The weight w_i^2 for the first interval estimation group can be calculated by Eq. 3. The comprehensive weight w_i can be calculated according to Eq.12 and Eq.13. The calculation result is shown in table 5.

Table 5. Weight calculation process and results

Index symbol	w_i^1	w_i^2	w_i
A_1	0.13	0.09	0.12

A_2	0.13	0.13	0.13
A_3	0.10	0.12	0.10
A_4	0.08	0.09	0.08
B_1	0.02	0.06	0.03
B_2	0.07	0.01	0.06
B_3	0.08	0.12	0.09
C_1	0.13	0.13	0.13
C_2	0.10	0.12	0.10
C_3	0.07	0.06	0.06
C_4	0.08	0.07	0.08

3.2. Evaluate the comprehensive benefit of biomass power generation project

(1) Determine the evaluation grey classifications, and select the threshold.

Evaluation indicators in this paper include qualitative and quantitative indicators simultaneously. In order to facilitate the calculation, qualitative and quantitative indicators are treated in the same way. Experts estimate the actual performance of the project, and the estimations are taken as the observation variable of the triangle whiten function.

Assuming that there are four grey classifications: excellent, good, medium and poor. The value range of them is shown in Table 6.

Table 6. The value range of excellent, good, medium and poor grey classification

Grey classification	Excellent	Good	Medium	Poor
Value range	[85,95]	[75,85)	[65,75)	[55,65)

(2) Optimize initial observation

Invite 5 experts to evaluate the performance of the biomass power generation project, obtain 5 sets of initial observation, and optimize the initial observation according to Eq. 16.

Table 7. The initial observation and optimized observation of evaluation Indicators

Index symbol	$x_j^{(1)}$	$x_j^{(2)}$	$x_j^{(3)}$	$x_j^{(4)}$	$x_j^{(5)}$	x_j
A_1	88	77	80	82	70	79
A_2	85	78	83	87	70	81
A_3	85	77	80	85	70	79
A_4	88	78	80	88	72	81
B_1	80	72	78	85	70	77
B_2	85	73	81	80	70	78
B_3	87	70	80	80	73	78
C_1	87	81	82	88	74	82
C_2	85	72	80	80	70	77
C_3	85	80	70	87	60	76
C_4	80	70	80	87	60	75

(3) Establish triangle whiten function

Extend the value range to $a_0 = 45$, $a_6 = 105$. Establish triangle whiten function based on table 6.

$$f_j^{\mathrm{P}}(x_j)=\begin{cases}0, x_j \notin [45,75] \\ \dfrac{x_j-45}{60-45}, x_j \in [45,60] \\ \dfrac{75-x_j}{75-60}, x_j \in [60,75]\end{cases} \tag{19}$$

$$f_j^{\mathrm{M}}(x_j)=\begin{cases}0, x_j \notin [55,85] \\ \dfrac{x_j-55}{70-55}, x_j \in [55,70] \\ \dfrac{85-x_j}{85-70}, x_j \in [70,85]\end{cases} \tag{20}$$

$$f_j^{\mathrm{G}}(x_j)=\begin{cases}0, x_j \notin [65,95] \\ \dfrac{x_j-65}{80-65}, x_j \in [65,80] \\ \dfrac{95-x_j}{95-80}, x_j \in [80,95]\end{cases} \tag{21}$$

$$f_j^{\mathrm{E}}(x_j)=\begin{cases}0, x_j \notin [75,105] \\ \dfrac{x_j-75}{90-75}, x_j \in [75,90] \\ \dfrac{105-x_j}{105-90}, x_j \in [90,105]\end{cases} \tag{22}$$

(4) Calculate the comprehensive clustering coefficient

Substitute the optimized observation in Table 7 into the above function to calculate the subjection degree of each indicator to each grey classification, and calculate the comprehensive clustering coefficient according to Eq.18. The calculation results are shown in Table 8.

Table 8. Comprehensive clustering coefficient calculation results

Index symbol	$f_j^{\mathrm{P}}(x_j)$	$f_j^{\mathrm{M}}(x_j)$	$f_j^{\mathrm{G}}(x_j)$	$f_j^{\mathrm{E}}(x_j)$
A_1	0.00	0.37	0.96	0.29
A_2	0.00	0.29	0.96	0.37
A_3	0.00	0.37	0.96	0.29
A_4	0.00	0.25	0.92	0.41
B_1	0.00	0.53	0.80	0.13
B_2	0.00	0.48	0.85	0.19
B_3	0.00	0.47	0.87	0.20
C_1	0.00	0.17	0.84	0.49
C_2	0.00	0.51	0.83	0.16
C_3	0.00	0.57	0.76	0.09
C_4	0.00	0.64	0.69	0.00
σ^k	0	3.31	7.01	2.06

As can be seen from the above table, "good" level has the highest comprehensive clustering coefficient, that is, the benefit evaluation result of this project is “good”.

4. Conclusion

(1) In this paper, the grey comprehensive evaluation index system of biomass power generation project is established from three aspects: economic benefit, environmental benefit and social benefit.

(2) In this paper, the grey comprehensive evaluation model based on triangle whiten function is improved by introducing group judgment to the weight calculation process and evaluation process, so as to improve the objectivity of the comprehensive evaluation result.

(3) The case study result shows that experts pay more attention to the economic and environmental benefit of biomass power generation projects, and for social benefit, increasing farmers’ income is more important. The comprehensive evaluation result of this target project is “good”. This example proves the feasibility of grey comprehensive evaluation method based on group judgment in the benefit evaluation of biomass power generation projects.

References

[1] Li Liangjie. Study on the sustainability assessment of biomass power generation projects [D]. Beijing: Beijing University of Chemical Technology, 2010.
[2] National Energy Administration. The 13th Five-Year Plan for Biomass Energy Development [Z]. 2016-10-28.
[3] Hu Liyue, Li Liang, Li Junsheng. Characteristics and environmental effects of biomass energy [J]. Energy and environment, 2012, (1): 30.
[4] Niu Zhiping. The sustainability assessment of rail transit projects [J]. Journal of Tsinghua University (Science and Technology), 2007, 47 (3): 319-322.
[5] Wang Yanjun. Comprehensive Evaluation on Guoneng Shangcai Biomass Power Project [D]. Beijing: North China Electric Power University, 2014.
[6] He Yongxiu. Comprehensive Evaluation Method of Electric Power and Its Application[M]. Beijing: China Electric Power Press, 2011.
[7] Dai Ran. The application of grey triangle whiten function in the post evaluation of hydraulic engineering project [J]. Harnessing the Huaihe River, 2009 (7): 22-24.
[8] Qing Song, Wang Ye, Peng Peng. Power System Security Evaluation Based on Grey Clustering Method[J]. Science and Technology in Industries, 2016(7):83-87.
[9] Liu Sifeng, Xie Naiming. A New Grey Evaluation Method Based on Improved Triangle Whitening Weight Function[J]. Journal of Systems Engineering, 2011, 26 (2):244-25.